Domestic Security Systems

To Margaret, for her patience.

Domestic Security Systems

A. L. Brown

Newnes

OXFORD AMSTERDAM BOSTON LONDON NEW YORK PARIS
SAN DIEGO SAN FRANCISCO SINGAPORE SYDNEY TOKYO

Newnes
An imprint of Elsevier Science
Linacre House, Jordan Hill, Oxford OX2 8DP
200 Wheeler Road, Burlington, MA 01803

First published 1997
Transferred to digital printing 2003

British Library Cataloguing in Publication Data
A catalogue record for this book is available from the British Library

Library of Congress Cataloguing in Publication Data
A catalogue record for this book is available from the Library of Congress

ISBN 0 7506 3235 6

For information on all Newnes publications
visit our website at www.newnespress.com

Contents

Preface

It has always been a necessity, in one way or another, to guard and protect ones' family and property. Today, with our modern consumer society churning out new technology goods at a frightening pace, the range of stolen items grows ever larger.

Many people have suffered the trauma of crime in their neighbourhoods and at the top of the list is burglary. It affects us all in some way or another, whether it is because our possessions have been stolen from us, or because we have to meet the increased cost of insurance as a result. This loss can also bring a sense of instability and uncertainty into the lives of those most affected.

To counteract this growing social problem, various means of protection have been devised over the years. However, the simple mechanical and electrical devices such as five-lever locks, door chains, optical spyholes and basic alarms have given way to a bewildering array of highly sophisticated electronic systems.

Once, such systems would have been used solely by the rich and famous, or by banks and art galleries. Now that the cost of these systems has fallen dramatically, due both to an upsurge in demand and technological advance, most people can either have them fitted, or install them themselves, at a reasonable cost.

The object of this book is to familiarize readers with their security systems, learn their operation and understand how they work. Thus, some of the mystique will be removed; they will seem less complicated and a little more user-friendly. This book will also help the reader to appraise the masses of commercial equipment appearing on the market.

For the determined DIY enthusiasts, there is also information on all aspects of electronic home security: alarm systems planning, component selection, installation, repairs, modifications and upgrades are fully discussed. Chapter One describes the many types of input sensors likely to be found in the home, how they function and how best to integrate them into your security system and the pitfalls to be avoided. The various control panels with their multitude of features and the many output signalling devices for calling the alarm are also described in their own respective sections.

The installation of security lighting for perimeter protection, including manually switched and automatic sensor-operated types, are shown together with suggested circuitry for the inclusion of low-power sounders to deter prowlers and alert the occupants.

A section dealing with electronic door entry equipment shows how the most basic devices can bar entry to unwelcome guests. The recent introduction of integral video monitoring has added a whole new dimension by giving the user a clear view of the caller – a veritable boon to those who have hearing or walking disabilities. The new domestic video camera surveillance systems are also addressed, describing the now readily available miniature cameras and associated control equipment developed specifically for home use. Fully integrated systems are now available, which will keep a sharp lookout for prowlers and automatically inform you of their presence or record their images and voices on a video recorder if you are not at home.

For those who enjoy constructing electronic devices there are several experimental circuits to try. There is no constructional information for these projects – the details have been left up to the constructor.

All or some of these systems can be combined to give you a home secure from unwanted callers at any time of day or night. Most of the equipment can be installed relatively easily by using the information given in this book coupled with the manufacturers' instructions. Most importantly, however, these systems can give you back the peace of mind that used to be taken for granted by most people.

A. L. Brown

It is assumed that the constructor or installer of the systems described in this book will have a basic understanding of electrical and electronic principles. The projects described are intended to aid understanding of how many commercial devices and systems work and to promote learning through experimentation. If any difficulty is encountered, then the advice or services of an experienced electronics constructor or fully qualified electrician should be enlisted. The effectiveness of any completed device or installation will always vary, dependent on its location and use, and will rely on the constructor or installer for its design and safety.

Chapter 1

Input sensors

At the forefront of any security system is a sensor that has been specifically chosen to detect the presence or the passage of an intruder. These devices work on various principles and it is the purpose of this chapter to guide the reader through the many modern types that he or she will be likely to encounter when installing, repairing or extending an alarm system. The correct selection and positioning of the sensors may be crucial in avoiding false alarms, while at the same time utilizing their best potentials. The typical construction and basic operation of the various types are described in sufficient detail to give the user an understanding of the theory involved and how each type can be chosen for the many different situations in and around the house.

Switch sensors

Magnetic reed switch 'contacts'

The commonest and simplest sensor of all is the humble reed switch. It will be found, paired with a matching permanent magnet, fitted mainly to doors and sometimes the opening parts of windows. It is used to detect whether the door or window is closed on setting the alarm and also to trigger the alarm if the doors are opened by an intruder.

The body is a glass tube with a pair of open contacts embedded in it. These contacts are made from magnetic wire inserted into each end of the tube so that there is an overlap in the middle. They are attracted together when they are in the presence of a strong magnetic field. Encapsulating the contacts within a glass tube keeps the contact surfaces clean and free from corrosion, thus increasing the life of the switch and reducing the likelihood of false alarms. The use of a magnetic field as a means of operation, instead of levers and buttons, lessens the chance of mechanical failure – a lifetime of several million operations is a real possibility! The operating distance varies slightly with the strength of the magnet, the sensitivity of the individual reed switch and the type of housing used. Generally, the contacts are attracted together when the magnet is at a distance of about 6 mm, but released when the gap increases to about 10 mm.

Three basic patterns are available from the manufacturers, with some variants. The selection of types shown in Figure 1.1 are intended for internal use. Designed for surface or flush fitting, their bodies are made

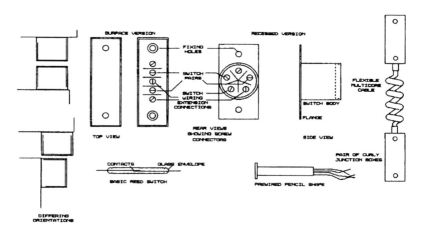

Figure 1.1 Assorted magnetic switch sensors

of plastic and can be coloured either white or brown in order to blend into the surroundings as much as possible. Connections may be by screw contacts or by soldering directly to the wires.

Figure 1.2 describes how the switch of the surface version is mounted on a fixed door frame with the cable connected to it. The paired magnet is fitted to the door in close proximity, with a gap of less than 5 mm between it and the contacts when the door is closed. If the door is opened, the distance is increased and the magnetic field is weakened. The demagnetized contacts lose their attraction and move apart, thus breaking the circuit.

In contrast, the flush fitting switch and magnet pair are set into the door and frame, making them almost invisible to casual inspection. This maintains the security of the alarm system by making the switches less vulnerable to physical attack. Figure 1.3 illustrates how a typical pair of these are fitted. Although the surface types can be fitted almost anywhere, the flush version, for example, cannot be set into plastic double-glazed doors or windows. This must be taken into account when purchasing parts.

If pre-wired switch contacts are used, then it might be necessary to use a purpose-designed junction box to connect them to the zone cable. Occasionally it may be desirable to mount a switch and magnet onto the two halves of a double door. Junction boxes with a four-core curly lead are available to allow the door to open without fear of the cable breaking.

A heavy-duty aluminium version is offered for use on garage or roller

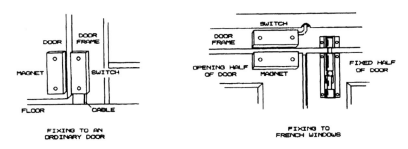

Figure 1.2 Positioning of surface switch sensors

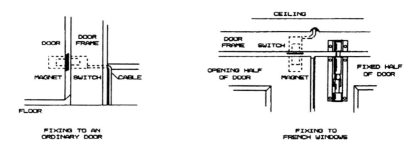

Figure 1.3 Positioning concealed switch sensors

doors. These are usually pre-wired with an armoured cable for protection against accidental damage. The switch body is often screwed directly onto the concrete floor of a garage, while the magnet would be bolted above it onto the moving door. Figure 1.4 shows how this is achieved for best effect and security.

Floor pressure mats

Hidden under carpets anywhere in the house, these mats can detect the pressure of a footstep. Several locations around the house are well suited for good security, such as near the television and video, music centre, or a collection of expensive ornaments. A smaller size will fit neatly

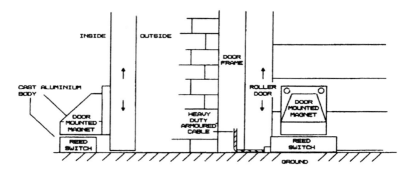

Figure 1.4 Heavy duty roller door switch sensor

on the stairs to detect anyone moving from one floor to another. Pressure mats should be considered as low security devices: because they are open circuit devices, a fault may not result in an alarm call and, without regular testing, a broken wire may lie undetected for some time. Only a short circuit will show up on the control panel as a fault.

Figure 1.5 shows the construction of a typical mat. Layers of plastic foam, aluminium foil and a protective PVC cover are combined to form a sandwich. The zone connections are made to wires connected to the foil sheets forming the normally open switch. When pressure is applied to the mat, the foil contacts are forced together through holes in the foam to make contact. The 'normally open circuit' nature of the mat switch contrasts with the 'normally closed circuit' of most other types of sensor, whether active or passive.

The mat is laid out underneath the carpet in the selected position. Its location is not betrayed due to its very thin construction. Carefully note where the mats have been placed, in case any heavy furniture in the room is rearranged. Otherwise, a fault condition may be generated that would require a great deal of work trying to clear.

Pressure mats are often used where the location or environment would not suit other sensors, or even places where their use would not be cost effective or practicable due to a restriction in the size of the area.

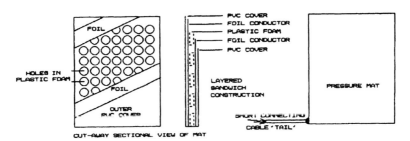

Figure 1.5　Pressure mat construction

Active sensors

Ultrasonic motion detection

Early motion detectors used reflected sound waves for protection against intruders entering a room. A generator produces a constant, stable, sound output to fill the room with inaudible waves. Sometimes these waves coincide with each other, either adding or subtracting from their nominal values. If there is no movement in the room then these combinations or 'standing waves' would be static, but if there is a disturbance then they vary in strength and vector. The direction, velocity and strength of these waves are dependent on the disturbing influence. There is a distinct similarity in this process to sea waves as they reach a breakwater. Moments after they hit the wall they are reflected back. Often, incoming and reflected waves will meet and clash; this is the equivalent of the sound waves combining. If the sea wall could move, then the waves would meet in a different position, depending on that movement. These then would be travelling 'standing' waves. The detector is thus required to be sensitive only to these moving waves. This type of sensor is thus termed 'active' because it generates its own detection medium.

A transmitter and receiver transducer pair are constructed using a special ceramic material possessing properties similar to that of a quartz crystal. The piezo ceramic deposited on to a brass or stainless steel disc will bend if a suitable voltage is applied to it. Just like a loudspeaker, a sound will be generated if an alternating (AC) signal is used. For ultrasonic operation, the frequencies used will need to be well above the threshold of human hearing. To have a useful wavelength, a 20 kHz operational frequency would be the minimum. The most popular frequency today is 40 kHz. The wavelength at 40 kHz can be determined by the following method:

> The speed of sound is about 1142 feet per second or 13 704 inches per second. Divide this by 40 000 (40 kHz, the frequency) and the answer will be a wavelength of 0.342 6 inches,
> or,
> wavelength = velocity / frequency.

This means that, for every third of an inch that a target moves towards the detector, a new wavefront will reach the receiver transducer. A series of these waves will appear as a low frequency alternating signal. In Figure 1.6 there is a block diagram of a typical ultrasonic sensor that will detect these waves.

For overall long-life stability, a quartz crystal oscillator is used to drive the bridge amplifier powering the transmitter transducer. The generator circuit shown is typical of these devices and will give a reasonable output and sound level using a standard 16 mm diameter transmitter sensor.

In this unit, the signal is first amplified and filtered before being shaped by a Schmitt trigger. A monostable is then employed to lengthen the output relay drive pulse sufficiently to trigger the alarm control unit correctly.

Figure 1.7 shows how these devices are constructed. Both of the sensors are mounted on the same circuit board in close proximity to the circuitry – the intention is to give a physical, as well as an electrical,

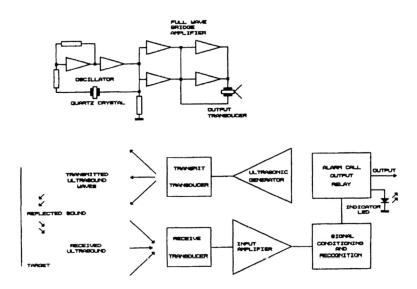

Figure 1.6 Ultrasonic motion detector (block diagram)

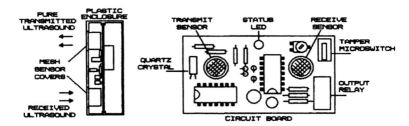

Figure 1.7 Ultrasonic motion detector construction

stability. An LED mounted on the board gives the status of the sensor by glowing when it detects movement. A typical operational sensing distance would be several metres in an enclosed area. If there is little chance of movement, then the sensitivity may be turned up to a value which stops short of generating false alarms.

Mounted in a suitable plastic enclosure, these devices would be fixed in the corner of a room. Any movement would then destabilize the sound waves, the maximum effect being in the direction to and from the sensor. Unfortunately, anything moving in the area could trigger these units, including curtains and even draughts of air!

Microwave motion sensor

This is also an active detector, but this time, instead of generating a steady, constant ultrasonic signal, radio waves are used to a much better effect.

The technology in this case is based on a special device known as a Gunn Diode. Selected for the frequency band in use, the diode is mounted at the closed end of a metallic cavity. This cavity is sized and shaped to suit the application and may be formed from cast aluminium or machined from a block of brass. Rectangular in section, (0.9 inches × 0.4 inches) the cavity is closed off at about ½ of a wavelength from the diode and, to assist in fine tuning, a screw is threaded into the end. The open end of the waveguide has a flared opening. This is the horn antenna: its length and flare angle determines the overall gain, and therefore the directivity and area coverage. Power for the Gunn Diode is a stable seven volts DC. Properly set up and tuned this will give an output in the 10.7 GHz band (1 GHz = 1000 MHz) set apart for the use of these detectors. The wavelength is three times longer than the equivalent ultrasonic unit at about 2.8 cm or 1.25 inches.

The receiver has a very similar arrangement. It too has a cavity and diode mounted in it. However, this is a special diode-type known as a mixer. As its name suggests, it is used for mixing a reflected portion of the transmitted output and the received signal. Being a non-linear device, this produces an output suitable for feeding directly into a detector circuit. The signals F1 (transmitted) and F2 (reflected) are combined together to produce two outputs. One of these is the sum of the two (somewhere in the region of 21.4 GHz), whereas the other signal is a stable beat note when there is no movement in the area. In a similar manner to the ultrasonic motion detector, low frequency waves set up by a moving object are detected and amplified and used to operate the output relay.

Sound, air movements and draughts have little, if any, effect, but other solid objects moving even beyond walls and floors can trigger the sensor. Microwaves are a form of radio wave and can penetrate plaster, glass and even brick, but they are much weaker as a result. To overcome these extended range problems a built-in sensitivity control can be adjusted to reduce the overall effectiveness, allowing only movement within the protected area to trigger the alarm. In Figure 1.8 the basic components are given, with a block diagram of the typical detector.

Different antennas can be employed to suit various room shapes and to give the best coverage, e.g. a long tapering shape for corridors or a more elliptical shape for a wider room.

Passive sensors

Electronic vibration sensor

These devices are intended for the detection of a forced entry by breaking through a window or door. The knocks and bangs inevitably involved generate sound peaks strong enough for such a detector to be triggered.

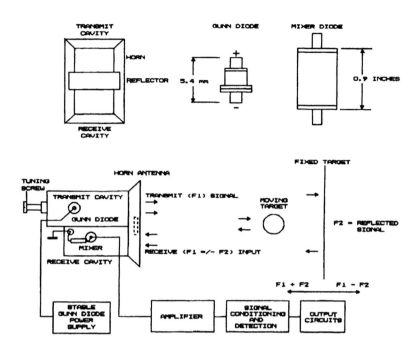

Figure 1.8 Typical microwave detector operating principles

Early vibration or shock detectors often used a glass bulb filled with mercury. Fused into the glass were two wire electrodes which would be covered with mercury when set at a specific angle. This angle was critical for sensitive operation, though it could be adjusted to suit the application. When a knock was applied to the window to which it was attached the mercury was momentarily displaced due to its inertia. For a short period the circuit would be broken as the mercury moved away from the wires. This brief open circuit would then be detected and the alarm sounded.

Modern electronic detectors are much more sensitive and reliable. There are virtually no moving parts to wear out and they can be adjusted through a very wide range in order to find the optimum sensitivity.

Quartz crystals have the property of generating a voltage whenever they are squeezed, bent or even compressed. One example is the ignition system in your gas fire or electronic cigarette lighter. When the button is pressed, a spring loaded bolt hits the crystal with considerable force – sufficient to induce a high enough voltage to create a spark several millimetres long.

In the vibration detector, a bar of quartz is fixed by one end, leaving the other free to move when subjected to any motion. This causes a minute voltage to be induced which is fed directly into an amplifier. The signal, which mimics the strength and frequency of the vibrations, is amplified, filtered and then conditioned to drive the output relay and illuminate a light emitting diode to indicate the sensor status.

When power is first applied to the sensor, the relay is energized, thus closing the alarm contacts. If the sensor receives a shock of sufficient strength it will deactivate the relay, breaking the circuit for long enough to trigger the alarm. Figure 1.9 shows a basic block diagram and how a typical modern vibration sensor is constructed. All of the components are mounted on a printed circuit board which is held securely in place by four plastic retaining clips. The gain adjustment control is a miniature horizontally mounted potentiometer. The quartz bar is semi-enclosed in a metal housing, both for physical protection and also to reduce the likelihood of radio frequency (RF) pick-up.

A microswitch set at one end of the printed circuit board (PCB) is used for anti-tamper detection. Should the sensor cover be removed, then the switch lever will be released, operating the switch and breaking the tamper circuit, thus resulting in an alarm call.

Passive infra-red sensor

This type of detector is without doubt the most common room sensor in use today. It can be found in both internal and external locations around many homes. It may be used in alarm systems, for switching on security lighting and even as a controller for your interior lighting in order to save energy.

Compared with the active types, this sensor is relatively simple and, other than the alarm output relay, it involves no moving parts and no generated signals. A specially developed pyroelectric detector is the

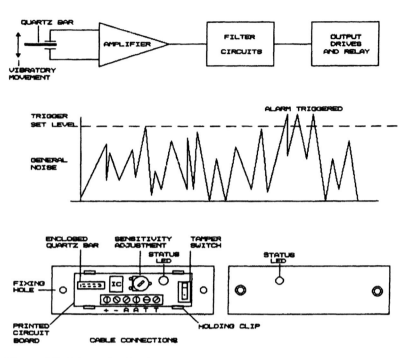

Figure 1.9 Electronic vibration detector

central component, which detects infra-red (IR) radiation generated by
a heat source and converts it into an electrical signal. If the heat source
is variable, then the signal output will reflect this. In Figure 1.10 it can
be seen how a series of lenses placed in front of the detector will focus
the radiation from the target area on to the sensitive element in the form
of 'beams'. Any moving heat source, such as an intruder walking across
the field of view, would cross these 'beams' and radiate sufficient IR
energy to stimulate the detector. Radiation would reach the detector in
a series of on–off pulses as the intruder's body entered or exited the
beams built up by the lens array. The resulting electronic signal would
then be processed into a form suitable for triggering the alarm.
Unfortunately, there is no discrimination between an almost static, but
fluctuating, radiation source and a moving intruder! To counteract this,

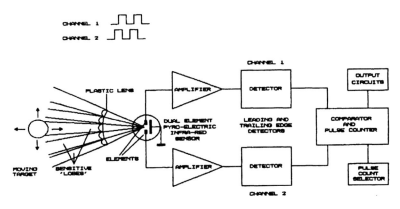

Figure 1.10 Basic infra-red sensor theory

a detector was developed which had two sensitive elements instead of one. This, and a partial duplication of the processing circuits to form a comparator, reduced the likelihood of the false alarm. Having a detector with two elements meant that an intruder would now have to travel across the area for the beams to focus the IR onto each element in turn so as to generate an alarm call. Discrimination between local heat sources is now possible.

In Figure 1.11 we can see how a typical modern IR room sensor is constructed. The plastic enclosure is designed with the back of the case shaped and angled to allow the sensor to be fitted neatly into a corner to give the maximum possible all round coverage. A PCB, populated

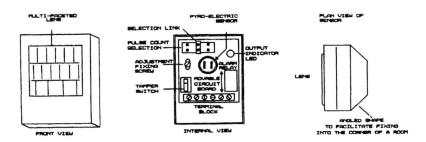

Figure 1.11 Typical infra-red sensor construction

with all of the electronic components, is slotted into a custom-made plastic housing. The board usually has some scope for vertical adjustment to control the height of the sensitive area. The front of the case has a semi-transparent plastic lens array to give the basic coverage characteristics. There are many types of lens patterns available, but the most common ones in use are given in Figure 1.12.

A volumetric lens is the most usual general-purpose type: three or more layers or rows of individual lenses are moulded into the plastic to cover both the horizontal and the vertical planes. Generally, the horizontal detection angle is 90 degrees but some, especially external ones, can be up to 270 degrees. The sensor so far described can only detect movement parallel to it, but, with the addition of the volumetric lens as shown, it can be seen that a warm object moving towards the sensor will cross the vertical beam layers, generating a similar signal pattern as before and thus triggering the detection circuits.

Other standard lenses include one which is intended to avoid detecting your pets. By only using one horizontal row of lenses, the plane of

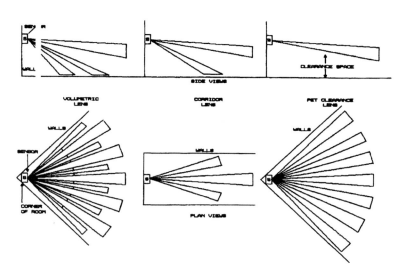

Figure 1.12 Typical infra-red lens patterns

detection is deliberately kept a few feet above floor level. This allows your cat or dog to walk free, while an intruder will cause an alarm.

Another, but less popular, lens is designed for use in a corridor. Although it is not commonly used in domestic situations, it can be found in larger houses or pubs which may have long corridors. The lens pattern is often confined to one large main beam accompanied by two or three smaller side beams.

The sun can cause real problems if care is not taken. Smooth shiny surfaces reflecting sunlight up to the sensor can have the same effect as a moving heat source, especially when there are fast moving clouds in the sky, which virtually switch on and off the sun's IR emissions. A similar situation is possible when doors or windows in the view of the sensor have convoluted or patterned glass. These act as a multitude of lenses and prisms, all reflecting, bending and splitting the sunlight. Sensors directed towards such doors and windows need to be desensitized to avoid trouble.

One of the more recent innovations has been the introduction of the semi-transparent white light filter lens. This lens reduces unwanted interference from the biggest IR producer around! By using a white plastic material, the lens allows a reduced amount of sunlight, whether direct or diffused, to reach the detector, thus reducing the possibility of false alarms.

Combination units

Combined IR and microwave detectors

In an attempt to realize the best features of both the IR and microwave sensors and to eliminate the various disadvantages and shortcomings of the individual detectors, a combination unit has been developed. This device merges the two technologies into a single sensor to give a much higher degree of reliability.

Activation of the alarm output requires a totally different pattern of events to occur in the detection area. To generate an alarm call, both sections of the unit must be triggered almost simultaneously. However, the diagram in Figure 1.13 shows not only how both sections can be activated separately by different events, but also how they must overlap to illuminate the output LED and energize the alarm relay.

This is an important step forward, because only a warm, moving, object can trigger the sensor output. Neither a flickering flame in the grate, nor the warm air blowing gently from a central heating grille will be quite so likely to cause an alarm. Similarly, a door suddenly blown open in a gust of wind or some other cool, but moving object is far less likely to be detected. An intruder, though, is very different: he or she is both warm and moving, and is therefore a genuine target.

On the down side, radio interference may trigger the sensor, but in these days of CE marking of equipment, even this event is becoming increasingly unlikely.

Combined IR and smoke detectors

Another combination unit becoming popular is the IR sensor with a built-in smoke detector. Not only will the device detect flames (IR), but smoke detection is now possible before any flames are produced. Correctly wired it can give the family 24-hour protection against fire

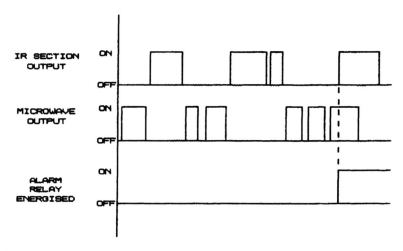

Figure 1.13   Infra-red and microwave unit outputs

or the effects of smoke. Effectively, this gives the standard domestic alarm an added dimension or purpose.

Smoke may be detected by one of two methods:

- **Ionization.** This method is similar to the stand-alone battery powered detectors found in millions of homes across the country. Smoke entering a semi-enclosed chamber is bombarded by a radioactive source causing it to be ionized. A potential difference is set up between a pair of electrodes in the chamber. When a large enough charge is detected the circuits activate the smoke alarm relay. These detectors are inexpensive and respond well to fires, but, because they can detect invisible particles of smoke in the air, they are less than suitable for the kitchen.
- **Optical.** An IR LED is used in this design to send pulses of invisible light into a darkened chamber. The IR detector is placed in such a manner that it cannot normally receive this radiation, but will pick up reflected light from any smoke particles entering the chamber. Again, sufficient smoke or fumes must be present to raise the detected level of IR to trigger the alarm.

The size and design of one of these units compares favourably with a typical domestic IR sensor so that it would not be out of place in any home or office. They cost more, but this is due to the increased complexity of their design. Fumes from smouldering fires are usually detected best by this type and they can be used everywhere, including in the kitchen.

Because they are in permanent use, these units are very useful in extending the capabilities of an existing alarm system. An early warning of fire or smoke can save lives, so at least one of each type should be installed as a good investment. Battery powered types may still be fitted in most rooms to give added protection.

Note. The use of these detectors should not be confused with a fully installed dedicated fire alarm. These systems are purpose built to exacting British Standards.

Miscellaneous switches

Personal attack button

There may be a time, whether by day or by night, when it becomes necessary to sound the alarm for personal protection. Strategically placed latching buttons can be employed to trigger the alarm and call for help. Figure 1.14 describes such a switch. They are either made from metal (industrial) or more likely an off-white plastic for the home. They all have one common feature, i.e. a large red button.

When pressed, the button moves a magnet away from a normally open reed switch in a manner similar to a switch contact pair fitted to, for example, a door. The button latches into this open position mechanically and can only be released by the turn of a key. There is some protection against accidental operation of the button by virtue of it being slightly recessed.

Several buttons can be wired in series, any one of which will have the desired effect. Common positions include the front door (to warn off unwanted visitors) or in the bedroom in case the sounds of an attempted break-in wake up the occupants.

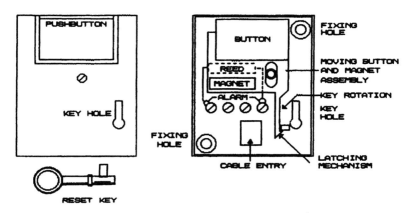

Figure 1.14 Typical personal attack button

Microswitch

Few of these switches are fitted to the modern alarm system. One remaining domain is in the external alarm box, as they are commonly used to prevent the box from being tampered with. Many alarm boxes have a base plate and clip-on lid with a single fixing bolt for holding it all in place. The microswitch is so positioned that when the bolt is screwed into place to tighten the lid it presses on and operates the switch. Any subsequent loosening of the screw will release the switch, break the circuit, and, depending on the system status, it will either sound the full alarm when the system is 'set' or the control box sounder if the alarm is 'unset'. This feature is known as the 24 or anti-tamper loop and will be discussed in greater detail in a later chapter.

Figure 1.15 shows the external structure of a typical microswitch. It has a sealed plastic body with two fixing holes moulded into it. A small button projects from one side of the switch body and is used to operate the internal mechanism. Only 0.3 mm of travel is needed to flip the contacts to make or break the circuit. This tiny amount of movement is far too small for the switch to be used successfully on its own. Because of this, a long metal lever-actuator pivoting in a recess in the switch body is used to increase the operating distance.

The life expectancy of the average microswitch is often in the order

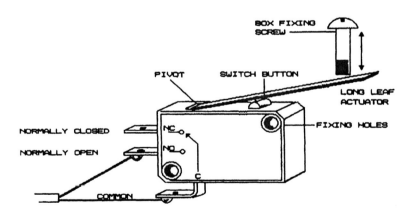

Figure 1.15 Standard pattern microswitch

of several million operations. The biggest threat is the ingress of damp-ness and the possible corrosion of the internal metals that go with this. Plating of the switch contact surfaces with precious metals goes a long way towards counteracting this problem, which will vary from one manufacturer to another.

Summary

All of the switches and sensors described have their own best and worst features. To help in their selection, location and the avoidance of possible problems, the chart in Figure 1.16 acts as a quick guide.

The most common types in regular use in homes today are the reed switch door contact sets and the infra-red sensor. Large areas can be

SENSOR TYPE	BEST USE IN	ACTIVATED BY	BEWARE OF
REED SWITCH 'CONTACTS'	DOORS AND OPENING WINDOWS	DOOR OR WINDOW OPENING	ENTRY VIA GLASS BREAKAGE
PRESSURE MAT	SMALL WELL DEFINED AREAS	DOWNWARD PRESSURE	FORGOTTEN POSITIONS
ULTRASONIC MOTION DETECTOR	ENCLOSED SPACES	ALL MOVING OBJECTS	HOT OR COLD AIR MOVEMENTS
MICROWAVE MOTION DETECTOR	VERY LARGE AREAS	MOST 'SOLID MOVING BODIES IN LOCAL AREA	EXTERNAL MOVEMENT TO LOCAL AREA
VIBRATION DETECTOR	DOORS AND FIXED WINDOWS	ATTEMPTED FORCED ENTRY	HEAVY RAIN OR HAILSTONES
PASSIVE INFRA-RED SENSOR	MOST LARGE AREAS	WARM MOVING BODY IN AREA	ALL HEAT SOURCES
COMBINED INFRA RED & MICROWAVE SENSOR	MOST LARGE AREAS	WARM MOVING BODY IN AREA	RADIO INTERFERENCE
COMBINED IR AND IONISATION SMOKE DETECTOR	LARGE FIRE-SENSITIVE AREAS	WARM MOVING BODY OR SMOKE	COOKING ON GRILL ETC.
COMBINED IR AND OPTICAL SMOKE DETECTOR	KITCHENS AND OTHER HIGH RISK AREAS	INTRUDERS, SMOKE AND FUMES	THICK TOBACCO SMOKE ETC.

Figure 1.16 The advantages and disadvantages of the different sensor types

covered by using more than one IR unit. Being passive, they will not cause any interaction between each other, but will provide a very high degree of detection, making it impossible for anyone to move in the area without being sensed. Mats are still used when a specific piece of floor area needs extra security. It may well be that the room is too small to warrant the use of a space sensor. Sometimes it is desirable to have mats fitted on the stairs leading to upper or lower levels in a house.

Electronic vibration sensors are often fitted to a window that is considered to be in a higher-risk location. This gives that extra bit of protection just where it is needed. Occasionally, the walls of a room may have a sensor fitted. This may seem strange at first, until it is realized that if the area is a store of expensive equipment or other items, it will need an increased protection level. After all, given half a chance, it would only take a few minutes for an intruder to break through a brick wall and take whatever is at hand. A vibration sensor would detect any hammering and possibly sound the alarm after only a couple of hammer blows.

The active sensor is not often used in homes these days, due to the upsurge in the popularity of the IR sensor. The ultrasonic motion detector is more likely to be found protecting the interior of a vehicle. Microwave detectors, whether as single or combined units, are much more common on the factory floor, where there may be heated machines. They are very suitable for most applications, but can be rather more expensive than the best PIR sensors. If a house has very large rooms, such detectors should be seriously considered as a good space protector.

Chapter **2**

Alarm system control

The complex digital control systems employed in the majority of homes today are the descendants of the early analogue devices that were popular for many years.

Analogue control

Before we can appreciate or understand the modern control panel it is perhaps desirable to learn the basic operations of the older versions. There are still many of these analogue control units in use and on sale, and many more still giving good service around the country.

A means is needed to enable the householder to set an alarm system, vacate the area, detect an intruder and sound the alarm and yet the occupier must also be able to re-enter the premises and unset or deactivate the control unit without tripping the system themselves.

The easiest way of achieving all the above aims is to use a series of timers. The first one needed would be the exit delay to allow a getaway from the area. A good 30 seconds should be more than enough for most of us. Once clear, you could relax, knowing that any intrusion into the secure area would trip the alarm.

So how do you get back to the control box without sounding the alarm and disturbing the neighbours? The answer is another timer, this time with a period of about 20 seconds to inhibit the sounders while the alarm control is switched off.

Now you have another problem: if you are able to get to the alarm box and switch it off then so can an intruder. If, however, the switch was key-operated and the key kept in a safe place, or on the person of the occupier, then a burglar would be unable to disable the alarm before the sirens were activated.

One more timer is now required to stop the alarm sounders after a nominal period. The maximum sounding time is 20 minutes, to limit the time the sirens are operating and reduce any unnecessary annoyance to neighbours if the system develops a fault.

Control panel basics

The diagram in Figure 2.1 shows the various building blocks for a basic control panel. First in line is a latching circuit – this is needed

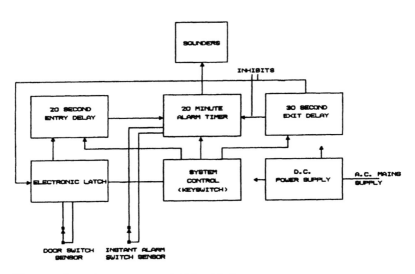

Figure 2.1 Basic alarm system building blocks

to hold the open-circuit switch condition when the entry door has been opened. This must still hold true even if the door is subsequently closed – without it, an intruder could close the door again and cancel the entry sequence.

Next is the entry delay timer – often adjustable from about 5–30 seconds to suit the user and the distance to the control box. In the control unit of an analogue unit the adjustment will be infinitely variable, but should it be a digital or push-button type it may only be changed in one second increments or more. This gives you the chance to reach the control box and deactivate it.

If the entry delay has expired without the system being unset, the alarm call timer will be enabled. This will drive the sirens until its delay period has 'timed-out' or has been disabled by the user. When the siren or bell has stopped sounding, the control will then automatically check the status of the switches. If all of the circuits are clear, then the alarm will reset itself and be ready for the next disturbance. If, however, a switch is still open then the sounders will restart.

At least one other circuit is usually available for input sensors; this is reserved for protection against entry by other routes – known as an instant alarm circuit or zone, it is often used on back doors or windows that are at greatest risk. The alarm has no delay in sounding when this circuit is activated.

When the system is first set or armed, an exit delay is immediately triggered – this circuit inhibits or puts a temporary hold on the entry and alarm call timers to stop the alarm from being sounded.

Wiring circuits containing sensors of any type are arranged into areas or zones. The correct planning and assessment of zonal areas will be fully considered in a later chapter.

An experimental alarm circuit

The circuit shown in Figure 2.2 has all of the basic needs of a simple alarm: it has the three timers we need – exit delay, entry delay and the alarm sounding timer. The integrated circuits used are the industry standard '555' timers configured in monostable or one-shot mode.

The exit delay is built around IC3 and transistor Q4. The alarm run

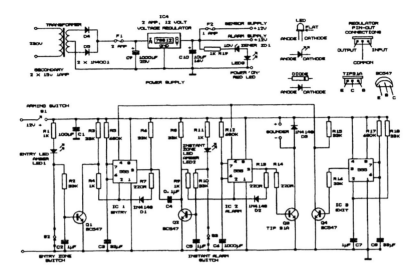

Figure 2.2 Simple analogue alarm circuit and power supply unit

timer and siren driver functions are fulfilled by IC2 and transistor Q3. Q1 and IC1 together perform the dual functions of input latch and entry timer. Q2 is used to detect the status of any switches in the instant alarm circuit.

When power is applied to the circuit via S1, trigger pin 2 of IC3 is momentarily pulled to a low potential by C7. This action forces the timer output, pin 3 of IC3, to switch high. The potential at pin 2 rises up above the trigger threshold and allows the timer to complete its cycle. Q4 is now turned on by the bias current supplied by R16 applied to its base connection. The reset pins (4) of IC1 and IC2 are pulled down to the negative rail by Q4, thus preventing the timers from being triggered.

When the exit delay has completed its timing period the reset pins of IC1 and IC2 are allowed to reach the positive rail potential via the common resistor R15. Both of these timers are now ready to be triggered if either S2 or S3 are opened. If S2 is opened, the base of Q1 is biased high, thus forcing its collector low. At this point pin 2 of IC1 is pulled below the 0.6 V trigger threshold. Output pin 3 now switches high and, after the entry period has lapsed, returns to a low level. The capacitor C4, which is now in a discharged state, pulls the trigger pin of IC2 low and recharges via R8. The output pin of IC2 now switches high and drives the base of Q3 which supplies the siren with power. Should the switch S3 be opened the entry timer is bypassed and the alarm is sounded immediately by pulling low the trigger pin of IC2.

In practice the resistors R5 and R16 can be a combination of a fixed resistance and an adjustable preset – these and the capacitors C3 and C8 are for setting the timing periods for the exit and entry delays.

The RC (resistor–capacitor) filters formed by R2 / C2 and R10 / C5 reduce any incoming interference that could trigger the timers. LEDs 1 and 2 confirm the status of switches S2 and S3: their lack of illumination indicates an open circuit and therefore a fault.

One problem remains with this particular circuit: the resistor R11 and capacitor C6 would have to be unrealistically large to allow the timer to run the full 20 minutes with any degree of accuracy. The capacitor may develop slight internal leakage – if so, then the timer may never reach its completion or may be erratic in operation.

As a rough guide to the RC timer values, the combination of 680 kΩ and 33 µF for R17 and C7 would give an approximate 33 second timing period. For a 20 second delay, use 680 kΩ and 22 µF for R5 and C3. If R12 and C6 were valued at 680 kΩ and 1000 µF then the alarm should run for about 10 minutes. These figures are dependent on component tolerances, which can vary considerably. In practice, the capacitors usually have actual values 20 per cent above the stated value. Occasionally, they can reach to 33 per cent above nominal, but this has to some extent been taken care of in the circuit design.

Power supply

A standard design 12 V DC stabilized unit capable of supplying 1 A is needed to power the alarm – the components must be purchased with this in mind and, by using the values given in the diagram, any shortfalls will be avoided. Fuses should be of the 'anti-surge' type to avoid them blowing at switch-on. Extreme caution must be exercised in the construction stage in the interests of safety. Alternatively, a ready made powerpack that can match the requirements may be used.

Alarm circuit add-ons

Extra features can be added to enhance the basic unit – these are intended to expand the capabilities of the system and make it viable as a practical and user-friendly alarm unit. Some of these are shown in Figure 2.3.

To verify that the exit delay is in progress, there is a simple LED add-on circuit: a resistor and LED may be connected between pin 3 of IC3 and ground. For confirmation of a valid power supply, connect a red LED and a 10 V Zener diode between the supply rails as shown. This, plus the 2 V to illuminate the LED, equals the 12 V needed for the supply output.

An entry zone warning sounder, to remind you the alarm is set and that it will need to be switched off quickly when you enter the house, can be added by connecting between the output (pin 3) of IC1 and the negative ground rail. Use a low power electronic type that uses only a few milliamps. By adding two diodes (D6 and D7), as shown, the buzzer can sound for both the exit delay and during the entry period.

Any number of extra zone circuits may be added by duplication and,

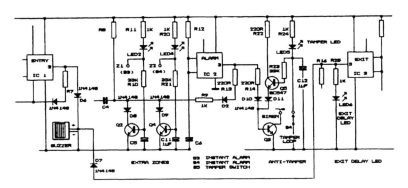

Figure 2.3 Alarm circuit add-ons and modifications

if necessary, one zone per sensor switch, etc., can be achieved. The zone diodes D8 and D9 stop any interaction between the zone circuits, but allow each to instantly activate the alarm. In addition, there is an extra supply output for powering up any active or passive sensors.

An anti-tamper circuit can be included, as shown, to protect the outside alarm box. The transistor Q5 is normally biased-off by the short circuit to ground on its base. If the circuit is broken, the cable cut, or the tamper switch opened, the bias to the transistor will rise to switch on the siren driver and the LED will be extinguished to show that the circuit is broken.

No PCB or construction details have been included for this circuit, but there is a parts list in the Appendix. The circuit would be suitable for a medium-level security system; it would be likely to cost only a few pounds and could probably lend itself to expansion and modification. There are no moving parts; no relays to wear out or stick, other than the mechanical on–off switch. Because of this, the unit should have a fault-free life span of several years.

Commercial analogue control

The block diagram in Figure 2.4 describes how the typical building

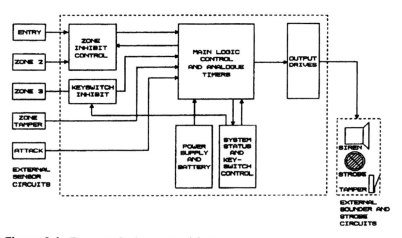

Figure 2.4 Four-zone basic commercial system

blocks of a commercial alarm control unit may appear. Although the system may seem to be more complex, it is, in fact, very similar to the previous example – additional circuits have been added to give extra protection, but basically the operation is the same.

One noticeable difference is the alarm inhibit which is now built into zone 2. This increases flexibility when considering the position of the control panel. Zone 2 is now what is commonly known as a 'walk-through' zone. This means that, if the entry zone (1) is activated, then zone 2 can be walked through to reach the control panel. If, however, zone 2 is entered without the entry zone being activated, then the alarm will be sounded immediately.

Zone 3 remains an instant alarm circuit, ready to activate the alarm if any intruder is detected. Sometimes this circuit may be switched off or disconnected if the panel is in 'part set' mode.

Part-setting an analogue panel often involves the omission of a zone by operating the panel key switch. If the system is installed in a house, then the occupants will need to have the upstairs omitted at night.

A personal attack circuit has been included in this example to give 24-hour protection. The latching buttons can be fitted almost anywhere, ready to sound the alarm at any time.

In addition to the siren, there is a separate output drive circuit – this time for a visual alarm. A flashing beacon utilizing a high power light is fixed to the outside of the alarm box. Whereas the siren will stop after about 20 minutes, the beacon will, with most panels, continue until the alarm is reset.

Digital control

By using digital circuit techniques, a system can be made which is much more complex, but which will offer a far greater degree of flexibility and a host of advanced features not found with the analogue types. Once the initial design has been proven, most of the circuits can be reduced to a single integrated microprocessor chip. This approach also increases reliability and reduces cost.

The main differences between the two systems are in the use of a clock generator, digital timing circuits that count the pulses, resetable latches to hold various conditions, and a memory to store user instructions or parameters.

In Figure 2.5 it can be seen how the design has changed. The addition of the clock generator as the heart of the system means that instructions and actions are carried out in a regular fashion. The timers are incremented at one-second intervals, making it possible to accurately set the delays, not by turning a screw on a preset resistor, but by entering the numerical values on an integral keypad.

To set most digital alarms, a personal code is entered on the keypad. This information is compared with the code held in the memory; if the two codes match, the exit delay will start to decrement in one second units. At the end of the period, when the counter reaches zero, the processor enables or 'arms' the system.

Now that the alarm is set or armed and ready to detect any unwanted intruders, it only needs a door to be opened for a fraction of a second for the input latch to close. If this occurs, an electronic gate, operated by a zone circuit latch, opens to allow clock pulses to reach the entry delay counter. Once started, this delay circuit cannot be stopped unless a reset signal is received from the central control. Entering the code at this stage will instruct the processor to reset all counters to zero. In

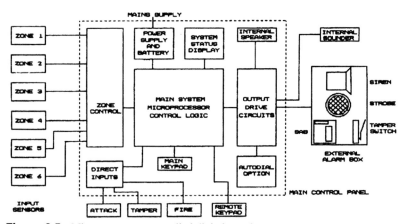

Figure 2.5   Microprocessor-controlled digital panel

this case it would stop the entry delay counter – if the alarm sirens were sounding, they too would be silenced.

Zone 1 is normally designated as being the entry and exit route. Zone 2 is sometimes switched off when zone 1 has been activated, but is otherwise an instant alarm zone. The remaining zones, usually numbered from 3 to 8, are, depending on the manufacturer, usually defined as instant alarm-only inputs. Some of the more recent panels allow every zone to be individually defined by the user – this increases flexibility at installation and allows for reprogramming at a later date.

In Figure 2.5, there are six detection zones, plus three other direct inputs. In practice, the number of zones available will vary from one type or model of panel to another. Four is the normal minimum, but some panels can have eight zones for sensor circuits. It is common to be able to control all of these inputs separately.

The direct inputs – attack and tamper – are not always accessed by the keyboard, but operate in the 24-hour mode for continuous, round-the-clock protection. The fire input is not a common feature, but it may be connected directly to smoke sensors or heat detectors designed for fire alarm panels.

The system status display, or annunciator panel, varies considerably from one type of alarm to another. Some may have single LEDs as visual indicators, while others may utilize a seven segment alpha-numerical readout to display numbers and letters. These latter may indicate zone numbers, error codes, fault codes, etc. Audible tones and beeps can also be used to convey information – they can also provide additional audio-feedback when the buttons are depressed. Tones are used for exit and entry delay warnings, unauthorized panel opening, panel alarm, zone recognition and walk tests, etc.

Extra remote keyboards can be added to some panels to provide more than one entry and exit route. All of the main control panel functions can usually be accessed from remote panels, including time delay settings, code changes, etc.

The minimum of two output circuits is normally provided: one is for a siren that will run for an average maximum of 20 minutes; the second is for a beacon strobe, which is often arranged to run continuously until the alarm is reset. Coupled to the siren output and tamper circuits we may find another device, though it is not strictly an output circuit or signalling device. The self-activating bell (SAB) is for added security at the external alarm box. It automatically provides power to the siren in the event of the wiring to the alarm box being cut. A rechargeable nickel cadmium battery, with associated electronic switching components, is mounted on a PCB, which is fitted into the alarm box. This constantly monitors the wiring between the panel and itself. Any fault will instantly result in an alarm condition, or at least cause the internal panel to sound a warning.

If an intruder were to locate and cut the cable leading to the box, the alarm will instantly sound. Even if the thief were to wrench the alarm box from its fixings, the siren or sounder will still be driven, even if it is carried away from the scene. Any internal sounders would also continue to deliver their shrill warning.

Control panel features

Control panels have a large variety of features: the style, shape and construction material will vary, as will the layout of the panel; the buttons may be of the older touch-sensitive type, or moulded rubber, or the newer illuminated plastic. Many of the more common, and some of the rarer, extended features are described below.

Day or unset mode

This is the normal condition that the alarm is in when not in use or deactivated. Some circuits will still work with the panel in this mode: these are the 24-hour zones, which include tamper, personal attack, fire, etc.

Set mode

The opposite to day mode, this is when the alarm is armed, the detector zones are connected through to the control and the system is readied to detect intruders. All input and monitoring circuits will now cause a full alarm if any interference is noted.

Zone omissions

When the alarm is set and certain zones are not needed at that specific time, then those zones can be omitted. Depending on the installation, zones in use in a bedroom or first floor landing area will not be required at bedtime. If a zone or zones are used for perimeter protection, then they can be used in isolation; ideal if the occupiers are watching television.

Intelligent auto-reset

This feature is part of the microprocessor control logic that is permanently written into the programming code governing the alarm. When a zone is activated and the alarm is triggered, the system sounds the alarm in the usual way. If any zone is still open when the sounders are reset, then the control processor treats this as a possible fault and disconnects that circuit. It may be an open door or a genuine sensor breakdown, but if the zone were not so isolated then the alarm would sound again until the occupier returned and reset the system. Unfortunately, this could happen when the occupants were away for a long period. This would probably upset the neighbours and cause unnecessary aggravation. Worst than this, it would reduce the validity and therefore the effectiveness of the alarm. With this feature, other working circuits would not be affected and would be ready to detect intruders.

User or customer code

The user or occupier will have his or her own code for operating the alarm. Four digits are common, but some may use eight; they may be of any desired combination that does not conflict with the system. Once chosen, the code can be kept for as long as it is thought to be secret, and if there is any doubt, then it can be changed – in fact, they can be changed as often as thought necessary to suit the user. If the house is temporarily vacated, e.g. for holidays, and someone is asked to come in from time to time, say to feed the fish or water the plants, then the code can be changed for that period only, so as to keep the normal code private.

This will then be reinstalled when the occupier returns, or changed if desired.

Test routines

Many control panels have preprogrammed routines that can be accessed from the keyboard. Zone tests may be carried out using the internal loudspeaker to give audible tones that are loud enough to be heard around the home during static walk tests. The siren and beacon strobe can be operated directly in this way, saving time in arming the alarm and then deliberately triggering the sounders.

Programmable delays

The delays for exit, alarm run and entry may be changed to suit the user or the building. In most systems, the changes can only be effected by entering the user and engineer codes. Without both codes, no such changes can be implemented.

Memory recall

If an alarm starts to exhibit a series of faults, it is helpful to be able to recall the recent fault history. Many panels can only indicate the last fault, but some can recall up to 16 past events. This information will tell the engineer if any one specific zone has more than its share of problems. If any maintenance is to be carried out, extra attention can be given to the suspect circuit.

Non-volatile memory

All programmed delays, codes, etc. are stored in memory. The older systems used volatile memories that, like computer RAM, lost all this information following a power cut. It was very frustrating for the occupier of a house to return to his home to find that he could not unset the alarm. It was also very embarrassing, because he would have to call the engineer for help and this could be at any time of day or night! The code (and all of the other preset information) would, in these circumstances, revert back to the original factory settings. Non-volatile memories changed all this, as the power could be cut off and the battery fully run down, but the code would remain the same on power up. Such memory can be provided by using EEPROM (electrically erasable and programmable read only memory), combined RAM and EEPROM on the same IC or the newer RAM chips with integral lithium battery.

Zone chimes

When an alarm is in day mode it is possible that an intruder could sneak into a house without alerting the occupants. The need for some sort of an alarm that will make anyone in the house aware of a door or window being opened or disturbed is a very real one. Many people have been burgled when they were watching television or had fallen asleep in a chair. The feature that can provide this protection is the zone chime. Few types of panel have this feature, but wherever it is included the zones can usually be individually programmed to respond, making the internal loudspeaker sound a distinctive tone. This should be sufficient to alert the occupier.

Cleaner code

For those who can afford to employ a cleaner to keep the home spick and span, but don't want the whole house to be available to prying eyes, some alarms can be set as normal when the house is vacated, but when the cleaner arrives and enters a special code, only part of the alarm is disconnected. The remaining zone or zones can be used to protect a room where expensive or private material is stored or used. The cleaner is thus allowed in the areas only where he or she should be.

The duress code and auto-dialler

If a family is unlucky enough to be burgled when they are asleep and the intruders have the audacity to drag the unfortunate sleeper from his or her bed and demand that they turn off the alarm, then the duress code can be used. If fitted, this feature silences the alarm siren, but what the intruder does not know is that the add-on telephone dialler has been activated and is sending out a prerecorded distress message. The recipient of the message can then call the police or perform a previously agreed action. The auto-dialler usually has the facility to call more than one number. The user should consider the likelihood of the call being answered and how the other person may react. One number that should be considered is the occupier's own mobile phone. If the house is left unattended for a short while and it is burgled, then the auto-dialler will alert the person who is going to be most interested and who can then decide whether or not to call the police.

Engineer code

When setting up or modifying the system, the engineer or installer will need to access the control via a specific route. Engineering programming involves the setting of the various time delays on a basic alarm and the advanced features, such as chimes or zone programming found on the more complex types. Entering the engineer code on its own should have no effect. It is common to enter the customer code first, as in this way it is deemed that permission has been obtained.

Fire zone

This zone, when included, is normally a 24-hour circuit. It can be used to call outwards if an auto-dialler is fitted, or simply sound the alarm. Smoke or carbon monoxide sensors, or the equivalent outputs from dual IR/smoke types are used exclusively. This ensures that only fire or the imminent threat of fire will activate this zone.

Individual zone tamper

Most alarm panels use only one tamper circuit to protect all of the sensors, zone wiring and the external alarm box. This makes it difficult to find where the trouble lies if a fault occurs. Some panels, however, have the added ability of being able to indicate which zone circuit has been interfered with, or has a fault. Tracing an open circuit break on a system with individual zone tamper circuits becomes much easier.

Personal attack

When the family are fast asleep and a noise wakes someone with a fright, a means to sound the alarm is a veritable boon. The personal attack button can be strategically placed so that it is accessible to as many of the family as possible. Pressing the button will sound an instant alarm. This should have two effects: it will awaken any sleeping children and/or partner; it will (hopefully) frighten away any prowlers before they actually manage to enter the house. Another use could be in the event of fire: if the system has not tripped due to detection of flames, the button may be pressed to help get everyone awake and out of the house. Neighbours will also hear the siren and hopefully give assistance.

Personal attack with silent alarm

The silent personal attack alarm will cause the auto-dialler to call for help. This would then allow an intruder to be caught unawares. Not only intruders can be removed – what about unwanted salespeople? Anyone causing an annoyance can be stopped if help arrives, and they would never know how they were frustrated.

Omit prevent

Normal omissions of zones allows an occupier to move around in specifically defined areas and keeps the alarm active in sensitive places. To fully protect a study or workroom and guard from accidental or deliberate omission of the zone, an omit prevent can be programmed into the system at setup. To remove the prevent it will be necessary to enter engineering programming and change the zone programming.

Wireless alarm systems

Hard wiring a conventional alarm can be a time consuming and dirty business. Wires are run to all of the sensors, under floors and in loft spaces; there is also the inevitable need to disturb furniture, carpets and sometimes whole rooms.

By contrast, the wireless alarm will often have only two cables: one is the connection between the control panel and the external alarm box; the other will be the mains supply. Instead of simple circuits with permanent wiring, which carry both power and signal, a coded radio link is established when the sensors are activated, to trigger the panel and the sounders.

To understand how this works, it is necessary to learn a little of the theory involved and how a digital identification tag can be impressed on to a radio signal.

Electromagnetic waves can be generated easily with a few simple components. The nominal frequency for alarm transmissions is 418 MHz. This is the same band used in car alarm remote control systems. Commercial systems must comply with a standard, which, for that frequency band, is known as MPT 1340. The maximum radiated power is only 250 MW, although this is more than sufficient for most home uses. Around 300 feet (91 m) from the control box is about the limit for sensor operation.

Figure 2.6 shows a typical circuit that can be used to create encoded

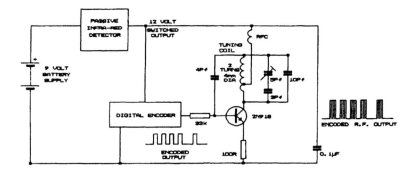

Figure 2.6   A sensor-operated encoded transmitter

VHF radio waves. The actual construction has not been described, but all of the necessary components and their typical values are shown. The tuning coil dimensions are critical – a slight variation in size can affect the output frequency by many megahertz. Often, the tuning coil (which, in this case, doubles as the antenna) is drawn directly as a PCB track for stability and a reduction in tuning drift and other problems associated with wire wound coils. The remaining components, including those for the sensor, can be mounted as normal on the PCB.

A digital encoder IC (there are many different types available) drives the oscillator transistor base with an on–off signal by switching the bias high and low with a series of pulses. These pulses form a code that is compatible only with the control panel zone it is associated with.

The sensor that triggers the radio transmission and therefore the alarm will be similar to those described in Chapter One. All of the components for IR detection, switch sensing, or even smoke detection are built with the encoder and transmitter as one compact unit. Automatic tamper protection is incorporated into the unit by using a switch behind the plastic cover. One other difference between these and ordinary systems is that every single sensor is self-contained and runs on its own internal battery.

Using one of these systems is straightforward: the user code procedure for exit delay initiation and entry delay switch off is the same as before, but with a wall-mounted wireless keypad. Another useful unit is a hand-operated remote control, which can access features such as instant remote on–off, full set, part set and personal attack, etc. all by pressing the appropriate button.

The external alarm box battery will probably be either mains or solar power charged. The mains version could be hard wired directly from the panel. The solar-powered unit usually relies on a coded radio signal from the panel to initiate the alarm sounding timer.

In Figure 2.7 there is a block diagram describing how such a system is built up. Each sensor function has its own code and these codes are set by adjusting a miniature switch array fitted on the PCB. This code must exactly match the code for the relevant zone and is unique to that panel. At the time of installation, the role of the sensor is decided, although this may be amended later if the system has a change of use or the home has a revised access route.

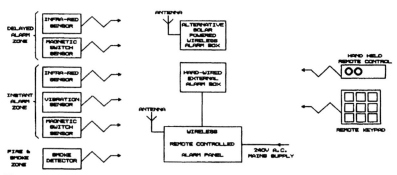

Figure 2.7 A typical wireless alarm system

The advantages of these alarms may not be too obvious at first, but it is a very versatile and flexible system. If the user is likely to move around the country as his or her job requires, then it may be ideal – it can be taken anywhere and fitted wherever it is needed in a fraction of the time of a conventional alarm. If the household budget is tight, then the components can be purchased separately, building up the system as finances allow. The sensors and keypad, etc. may be moved around to find the best coverage possible.

The main disadvantage is in the initial cost of the component parts. Saving on the cost of employing an alarm installer by fitting the system yourself could probably offset this expense and, of course, a removable alarm could then pay for itself many times over. Another additional expense is the need to change the sensor batteries. Twelve months is the average life expectancy of alkaline cells used in this situation.

Chapter 3

Output signalling devices

Signalling devices must attract the attention of yourself and family, your neighbours or the police. There are several gadgets and add-on devices that can make quite a lot of noise and there are other ways of making yourself or other people aware of a break-in.

DC bell

This must be one of the oldest types of electrically driven sounder and was used on alarm systems and doorbells. It has two moving parts: a vibrating bell dome for generating the ringing sound, and the hammer, which strikes it.

An ordinary hand-held bell is struck with single blows from a pendulum hammer hung from inside the dome. It produces a relatively short-lived pulsating tone that decays rapidly. An improvement to this was the clockwork-driven hammer, which struck the dome many times in rapid succession. It was very popular as a door bell, because it worked as long as the caller pressed the operating button, assuming it had been wound up in the first place!

Then came the electrically driven hammer of the standard DC bell, which also worked as long as a button was pressed.

Figure 3.1 shows two variants of a DC bell – the standard and the under-dome. The latter operates on the same basic principle as the original, but with a rearranged configuration to keep all of the workings hidden out of sight.

When power is applied to the terminals of the standard bell, the coils wound around the yoke are magnetized. Magnetic flux is concentrated in the U-shaped soft iron core. This magnetic field attracts the part of the hammer known as the armature. Attached to this is a light brass spring contact. The current energizing the yoke coils passes through this contact, but, as the armature is pulled towards the yoke, the spring leaves the point contact of an adjustable screw. At this point the current is interrupted and the magnetic field collapses. The whole of the armature is suspended on a steel strip spring which pulls it back so that the contacts touch once more. Again, the yoke is energized and the armature is attracted towards it. This cycle is repeated again and again until the supply is cut off.

The hammer at the end of the armature strikes the bell dome when the armature is pulled towards the yoke. The sound generated by the dome relies on how the hammer hits it. The adjusting screw can be turned until the loudest and most pleasing tone is obtained. A wrongly set adjustment can cause a clattering sound at one end of the range or a light tinkle at the other. Turning the dome may also give a small degree

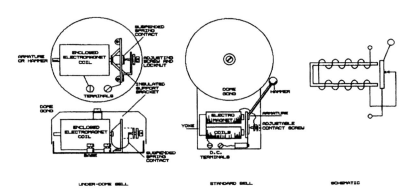

Figure 3.1 Two common DC bell types

of change in tone. Once set, the sound should remain satisfactory, at least until the mechanical components wear or the springs tire, when a slight readjustment will be required.

The average output of a 9–10 inch (225–250 mm) dome bell is around 96–104 decibels (dB). The sound itself may well be melodic, but this is not very loud. With the ringing at a constant volume, it may not be easy to distinguish a bell among other sounds that may be above normal levels.

The main disadvantage of using a bell for alarm purposes is the fact that the mechanical parts wear or become rusty and corroded when used outside. It may then need constant repair and adjustment to maintain the output level. The newer electronic types have replaced the mechanical contacts with a solid state circuit and are more reliable, but the ringing is still no louder.

Piezo electric sounder

A solid state sounder using only one moving part (the diaphragm) can be bought for only a few pounds. The sound level is often in the order of 104–110 dB. Current consumption is low, usually in the range of 85–190 mA. Figure 3.2 shows how a typical sounder may appear. The enclosure is normally made from ABS plastic, which is a strong but light material. The colouring of the plastic is a light cream or white to help it blend into the surroundings. It is designed primarily for use inside a building, where the humidity is likely to be low. Another reason for having a sounder inside the home is to create a 'wall of sound' that will disorientate, frighten or cause panic in the mind of an intruder.

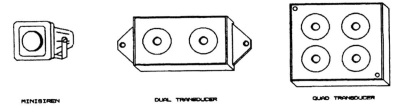

MINISIREN   DUAL TRANSDUCER   QUAD TRANSDUCER

Figure 3.2 Commercial piezo electric sounders

The sound is generated by the use of a ceramic material that has piezo electric properties similar to that of quartz crystal. A thin layer of ceramic (about 100 microns) is deposited on a brass or stainless steel disc (of a similar thickness), which is held in a resonant cavity. Applying a voltage to the transducer (the disc and the ceramic layer combination) causes it to bend. The direction and amount of bending is dependent on the applied voltage. DC current flow is zero, but AC is also low due to the device being mainly capacitive and of a very high impedance.

The cavity is a precisely calculated method of mounting a ceramic resonator. Three main factors govern the output level: the frequency of the transducer; cavity resonance; and the mass of the casing in which the sounder is fitted.

Some practical circuits, which can be used to drive one of these ceramic resonators, are shown in Figure 3.3. It must be remembered that, because the device will produce the loudest sound level at its resonant frequency,

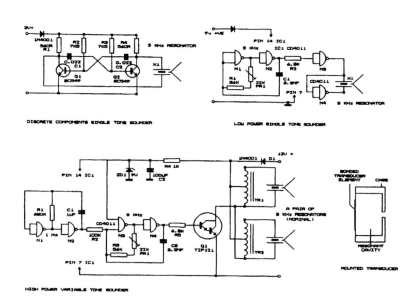

Figure 3.3 Piezo ceramic sounder circuits

any driving circuits must be designed to match this condition for best output. The simplest discrete components circuit uses two transistors in the standard multivibrator configuration. The transducer is connected between the collector output of each transistor. When each transistor switches on in turn, the transducer is energized first in one polarity and then in the other. This gives a peak-to-peak square wave drive of 6 V. The current, however, is limited by the 560R resistors in the two collector circuits. Another circuit using a CD4011 CMOS IC generates 24 V peak to peak (PP). It has a smaller component count and is easily adjusted for frequency trimming. N1 and N2 form the oscillator with the 56K resistor, 22K preset and the 3.3 nF capacitor. Both of the circuits are designed to oscillate around 3 kHz, which is a common resonant frequency.

In contrast to these two single frequency drivers, there is also a high output dual frequency sounder circuit. This uses two oscillators: one is tuned to the transducer frequency and the other is set at 1 Hz. The effect of the 1 Hz oscillator is to change the frequency of the second generator slightly. This is accomplished by adding or subtracting a resistor in the timing circuit and therefore changing the capacitor charging current.

By having a varying frequency output, the sounder is more easily heard above other noises. It is therefore much more penetrating and distinguishable as an alarm sounder. The output circuit is different – a coil has been inserted across the transducer terminals. This increases the drive voltage due to the voltage induced in the windings when the magnetic field collapses. This arrangement allows a much higher sound level from a single-ended amplifier circuit.

Another simple answer to increasing the sound level is to add another transducer (and transformer) to the transistor collector. This should double the volume for this type of sounder. Adding another two transducers will double yet again the output, this time to 110 dB. The power output has therefore been increased by 6 dB, or by 100 per cent. At close proximity, these units can really hurt your hearing, so wear some sort of protection when testing the alarm system.

External electronic siren

Although sirens have oscillator circuits that are very similar to those in the piezo sounders described above, the output is much louder, from 114 dB for a cheap unit to an earsplitting 130 dB with an expensive model.

Sealed plastic encapsulation of all parts virtually guarantees that no dampness can penetrate the electronics. Figure 3.4 shows the outline of some typical sirens and includes a cut-away sectional view of a popular style. The case itself contains the electronics to drive the cone-shaped voice coil. Backward-moving sound is reflected forward from the horn to increase the output level in much the same way as a voice megaphone directs the sound or a radio dish antenna focuses radio waves. This gives directivity and allows the siren to be pointed in whichever direction it would be most effective.

Instead of a piezo ceramic transducer, most sirens, but not all, use a small but high power form of electromagnetic loudspeaker. The vibrating cone is enclosed in a small resonant cavity to increase the sound pressure level. A small horn projects outwards towards a reflective cone which directs the sound back into the main horn. This results in a relatively compact but efficient construction.

Figure 3.5 describes some siren oscillators with a driver circuit. All three are open to experimentation. Using 555 timers, they are simple to build but have flexibility of design. They are all operated in astable oscillator mode. The second IC is set for a nominal frequency of 3 kHz,

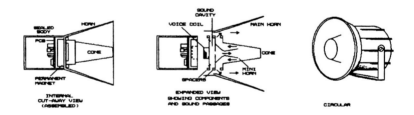

Figure 3.4 Typical siren construction

but the first IC of any pair is used to modulate this with a second, lower frequency, signal.

The pulsed tone generator is reminiscent of the early car alarm, where the car horn was used to sound the alarm. The horn was switched on and off, partly to give a distinctive, penetrating sound and also to save the horn itself from damage. The first IC simply resets the second by pulling the reset (pin 3) low to ground every half cycle. The components listed will give a 1 Hz pulse, but the 1 µF capacitor can be changed to either increase or decrease the pulse rate.

The pulse tone generator is very similar, but the modulating pulses are fed into pin 5, which controls the internal operating threshold of the second oscillator. Instead of the output being interrupted, the frequency is varied from one note to another. A 10K preset resistor can be adjusted to change the output frequency variation.

An adjustable sweep frequency with a slow attack and fast decay waveform again drives the control pin 5. The resistor, diode and capacitor arrangement between the two ICs controls the frequency of the second

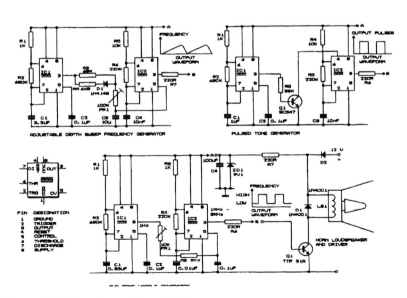

Figure 3.5 Experimental siren circuits

oscillator. When the output pin 3 switches high, the 10 µF capacitor charges up at a rate dependent on the combined value of the 68K fixed and preset resistors. Pin 3 switches low every half cycle and discharges the capacitor at a faster rate, giving a waveform similar to that shown.

The output driver stage is simple but can deliver the power needed to drive a 15 W horn speaker. The diode protects against reverse voltages generated by the coil. A Zener diode, a 100 µF capacitor and a 220R resistor stabilize the supply for the oscillator pair. Heavy duty 15–20 W horn speakers can be used, if suitably protected from the weather, with reduced output.

Flashing xenon beacon strobe

In order to reduce help response time and allow neighbours to determine immediately from where an alarm sound is originating, a visual indicator is needed. A xenon beacon, which flashes a brilliant light with a regular pulse, is ideal. These strobes have a nominal flash rate of 60–120 per minute. The colour range includes red, blue, yellow and clear.

In Figure 3.6 a block diagram describes the operation of a xenon flash tube. A high DC voltage of around 500 V is generated across the ends of the flash tube. In the low voltage types (12 or 24 V) this is accomplished by a simple transistor invertor. The energy, stored in a reservoir capacitor, is only about 5 Joules. One Joule is equivalent to one watt per second. This energy is discharged in a fraction of a second

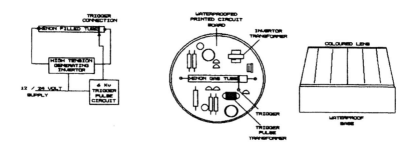

Figure 3.6 Xenon flashing beacon

when a very high voltage of about 6 kV is applied to a trigger connection near to one end of the tube.

The electronics draw a nominal 150–250 mA of load from a 12 or 24V supply. Normally, these units will have very well protected circuit boards, as they are subjected to whatever conditions the weather can summon up. Even if dampness does manage to penetrate through the case or coloured lens, it would still have to seep through a thick film of waterproof lacquer or similar protective coating. Sometimes, a high power 240 V beacon may be used, and this would require a mains supply and a switching relay.

These units are often mounted on the front surface of the external bell or alarm box. This ensures that the sound of the siren and the flashing light are readily recognized to be coming from the same location. The flash itself is often visible from quite some distance away, depending on background light and physical obstructions. An average lifespan may run to several million flashes – enough for a lifetime of alarm calls, false or not.

Alarm box

A housing is needed to protect and cover the siren from the burglar and the weather. The box shown in Figure 3.7 is a common example. Its sides, top and bottom are angled to reduce windage. Colours commonly available are blue, red, yellow and white. Although the colour can be picked according to taste, it should be remembered that the action of direct sunlight will, over a period of time, bleach the pigment, turning the box a lighter shade. Eventually, it can become almost white. For this reason, buy a box with a white cover if it is to be situated in direct sunlight.

Many of the older alarm boxes were made from mild steel. When they rust and the paint starts to peel, it becomes obvious to anyone, including prospective burglars, that the system is old and may be inefficient, or possibly not even working properly. A burglar would certainly give this house a try before attempting a break-in on a home with a new box, where the system could well be very sophisticated.

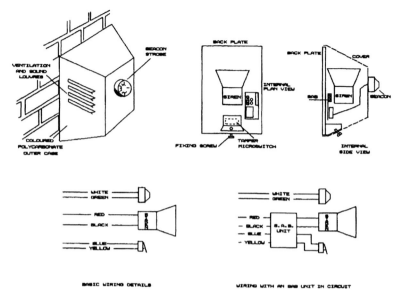

Figure 3.7 Typical alarm box assembly

SAB

Sometimes, a unit is fitted inside the alarm box that will sound the siren if the main cable to the box is cut. The self-activating bell has an integral rechargeable battery and LED charge indicator. Its circuitry ensures that, if any of the cable cores to the box are cut, the siren will be activated. The battery will power the siren until exhausted. However, the SAB may have a built-in electronic timer to switch off the siren after a 20-minute sounding period.

A microswitch is fitted in such a position as to sense the presence of the cover fixing bolt. A long actuating lever is pressed when the bolt is screwed into place, thus operating the switch. Should the bolt start to be unscrewed, the alarm will sound long before the bolt can be removed. Ready-assembled alarm boxes are widely available and can save much time in installation. On the other hand, a box built from component parts can save on the cost of materials.

If the alarm box fixing screw is tampered with, or the box is levered from its wall mounting and the wires cut, then the prospective burglar would either have to try and throw it away, smash it up, or drop it and run. The neighbours should by then have heard the alarm and the burglar would be likely to find that running away is the easiest option.

Dummy boxes can be fitted at the opposite side of the house to where the main operational box is situated. This can, simply by virtue of its presence, warn of the existence of an alarm, or it can be used as a pretence. Remember to match the type and colour of the dummy to the operational box.

Auto-dialler

If an intruder decides to enter your house and you live in the countryside or in a quiet area, where there are few neighbours or friends, can you be sure anyone will notice the alarm siren? You can if you had the foresight to have an auto-dialler fitted.

Early auto-dial units used a tape drive to play a prerecorded message. Because of this, they tended to be very limited and could often only dial out to one number, probably to the company that installed the system. A contract for keeping a watch on your property and for calling the police would have been signed; if your house had been burgled and one of these units were fitted, then the first you would have known about it would have been when you returned home to see the police at your door.

Modern units are a quantum leap forward: they are fully solid state with no moving parts; the dialling is performed by a single integrated circuit using tones instead of pulses.

Usually, up to four messages can be stored in a non-volatile memory IC. Sometimes, phrases can be used to build up the full message, which can be varied to suit the called number. A message to the police will be rather formal and will include the name of the occupier and the address of the property. Alternatively, the name and number of a contact nominated as a keyholder can be given.

A less formal message can be used for yourself or family. It would be simple and straight to the point. As mentioned in Chapter Two, your mobile telephone should be the first number to be dialled. You can then assess the situation and make the decision on whether to return home immediately or call for help.

If a smoke alarm is fitted in the system, or as a stand-alone unit, then it can be used to dial out in the event of fire. In this case, the message content would allow the fire brigade to be called.

When a message is received, the auto-dialler may need an acknowledgement of some sort. The act of picking up the receiver may be sufficient; if the call is not acknowledged, then the auto-dialler will call the next number on its list until there is an answer. These are invaluable for use in neighbourhood watch schemes, where the first number called could be the co-ordinator.

The messages to be relayed are recorded by the user. The length of the message is, however, limited and should be kept clear and concise.

Add-on devices

Other external devices can be connected to the alarm output. These extra features, shown in Figure 3.8, will enhance the system as a whole. Both low power and mains switching can be achieved by using relays. With the need for safety, separate relays should be used to switch the two very different voltage levels. This will prevent accidental damage to the alarm system and give some protection against shocks, should a mains wire touch part of the low voltage section.

Extra sounders

The first circuit shows how the relay can switch in an extra 12 V sounder. This method prevents the panel alarm output drive transistor from becoming overloaded if the siren has simply been connected in parallel with another. It is likely that a separate DC supply for the extra sounders would also be needed in case the internal supply and batteries were unable to cope with the additional demand.

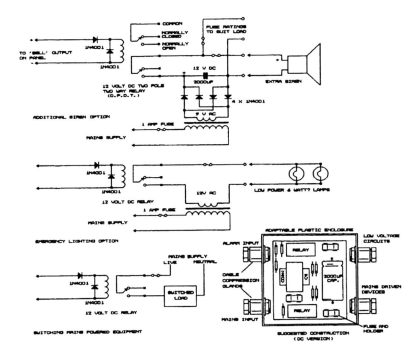

Figure 3.8 Alarm interfacing to add-on equipment

Emergency lighting

Not to be confused with fire alarm systems and dedicated emergency lighting, the control panel could be employed to switch on strategically placed low voltage or mains lighting. If the lamps are low voltage DC, then they can be run from a battery supply even during a power cut. This would be a boon should a fire start in the house. The exit route(s) would then be illuminated to show the way to safety. For this option, the power supply must have adequate capacity to cope with the load, which will increase with the number of lamps in the circuit and their rating. A fuse should be fitted in the low voltage wiring in case of short circuits or overloading.

External flood lighting

Although this subject will be discussed in greater depth later, it should be mentioned that, if any external security lights can be operated during an alarm call, then the combination of sounders and sudden illumination will be a greater deterrent to even the more determined thief.

Security cameras and recording equipment

This facility could be installed to automatically record any events, internal or external, starting from when the alarm was initiated. Again, the topic will be discussed later. The type of video system used will determine how it is operated.

Interface construction

Any interface must be constructed and fitted into an insulated plastic enclosure for protection against earth faults and the associated risk of shock. Fuses must be inserted at some point into each of the switched lines. The sketch shows a suggested layout. The alarm panel output must be connected to the circuit with the correct supply polarity, otherwise the relays will not operate. Without the diode protection provided, this could either result in a panel fuse rupturing, or the demise of the drive transistor. Care must be taken when wiring to mains equipment – if in doubt, ask for advice or call for the assistance of a qualified electrician.

Chapter 4

Installation

Installing an alarm system can be time consuming. How long it takes depends on planning, familiarity with tools and their use and the installer's experience. Those who have had little practice should take their time, plan the job, think out the next action, then carry it out one step at a time.

Planning the installation

To plan a burglar alarm system requires a good hard look at the home, its surroundings, gardens and outbuildings. Imagine yourself as a burglar. Can you see any weak points? Is it possible to visualize someone entering your home by forcing a small, half-forgotten window? Is there a partly hidden drainpipe that can be climbed? What about that flat roof above the garage, where a bedroom or landing window has been left vulnerable.

In Figures 4.1 and 4.2 are front views of a typical three-bedroom house and bungalow. In them are shown several weak points: with the house, there are those garage side windows and the one at the top of the stairs above the garage roof; those bushes at the side of the bay window could provide cover for a prowler; the sloping bay-window roof could put the

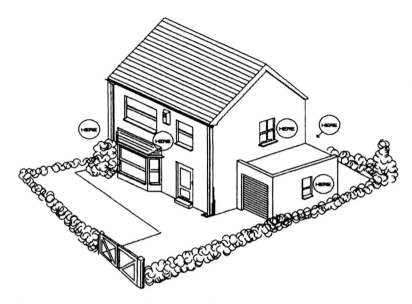

Figure 4.1 A typical detached house and its most vulnerable points

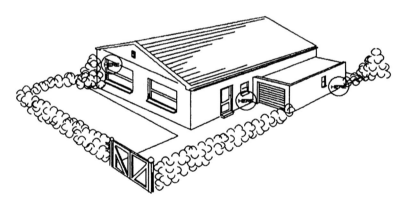

Figure 4.2 Our typical bungalow

front bedroom window at risk. At the back, out of sight, are two other weak points: patio doors and the rear exits from the garage.

In the case of the bungalow, the most vulnerable points are almost the same, although, obviously, all of the bungalow's entry points are at ground level. Not only small windows, but even picture windows are at risk. One woman didn't realize her kitchen window had gone during the night until she saw it lying on her lawn. Another incident involved a 10 foot wide by 5 foot high picture window that was removed so that the intruders only had to reach through the hole and steal the television and video. Later, three strong men struggled to replace the virtually undamaged glazing unit back in its original position.

Start by deciding on the basic minimum level of protection required and build up the system until you are satisfied that no point can be accessed easily and that the house is as secure as you can make it. Three levels of security are described below and the reader can judge how home and circumstances will best fit the designs shown.

Minimum security level

A plan of the ground floor of a typical house is shown in Figure 4.3. It shows likely locations of possessions in and around the building. It is the concentration of goods in the various areas that will effect the planning and installation of the alarm.

Top of the list for a quick-grab theft are the television and video, hi-fi and other entertainment equipment, usually kept in the living room. Jewellery, porcelain, ceramics and other artefacts come a close second with the planned intrusion. The increased popularity of computer equipment in the home has produced a new target area.

Whatever you do, don't underestimate the importance of security coverage in the garage, lean-to, or shed and the belongings kept there. Tools and equipment may be used to enter your home and may end up at the local car-boot sale, where there are few questions asked of ownership.

The second plan, in Figure 4.4, depicts the positions of the various sensors, in a low security design for a house. A plan for a bungalow is given in Figure 4.5. In these are shown the ground floor rooms, internal and external doors, the ground floor windows and their protection. At greatest risk of a forced entry are the patio and back doors and the back and side windows, which are out of view from the main road.

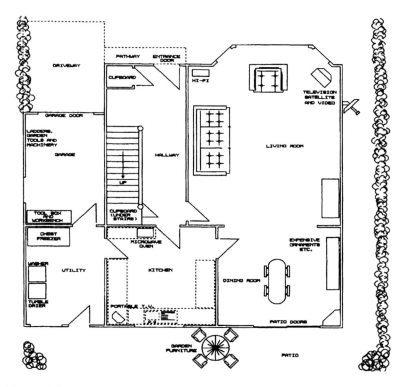

Figure 4.3 A typical detached house with examples of security risk items

The sensors shown in these plans have an emphasis on enhancing the main living area protection. The external and internal ground floor doors each have magnetic switch sensors and there is a passive IR sensor in the living room to detect any movement in that area.

The control box has been fitted in the cupboard under the stairs. This gives it some protection, by keeping it out of view from prying eyes. If it had been situated near to the front door, its position could have been noted by a caller who may have been checking out homes for his next job. If the control box is fitted into a cupboard or similar, it will be harder to find. Don't forget to secure any windows or other external openings that lead into the cupboard.

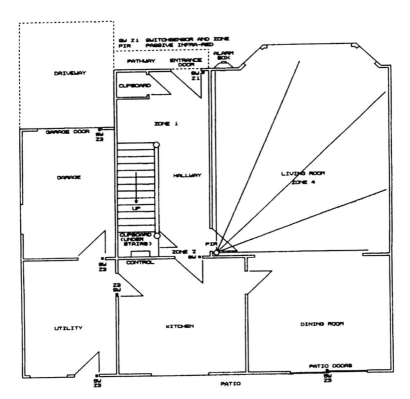

Figure 4.4 Low security alarm system (house example)

When all of the weak points have been identified, draw a plan of your home and then decide how you will exit and re-enter the house. Whatever is decided, the door, or doors, involved must be on zone 1. Most alarm panels use this zone to start the entry timer. Another basic decision to be made is how to get to the panel to turn off the alarm in the morning. In the given design, you can proceed directly to the cupboard where the panel is situated without triggering the system.

To give basic cover, four is a practical number of zones to create. This will give greater control flexibility and make it easier to find the source of false alarms. By dividing the sensors between the zones, such as keeping those in the entry route on zone 1, internal doors on zone 2, garage and external doors on zone 3 and the living room sensor on the zone 4, any fault can be narrowed down to only one or two positions. The ability of the control system to omit, or bypass, a zone then gives the flexibility of using only those zones needed at the time. If both zones 2 and 4 were bypassed or omitted in our house example, then the occupier could watch television, listen to music or even doze off, knowing that, if anyone tried to enter the home through one of the doors, then the alarm will sound.

On most modern control systems, zone 2 has a special characteristic: it is known as a walk-through zone. This means, quite literally, that the zone can be disturbed without sounding the alarm, but only when the entry zone has been activated and the entry timer is in operation. This is why you will find it used in an internal area, such as the kitchen door in our plans. It must not be used for the protection of high risk areas, because a burglar could then, in theory, enter the front door, start the entry delay, rush into the unprotected living room and leave with as much as can be carried within the 20 or so seconds before the alarm could be raised.

Zones 3 and 4 are normally instant alarm circuits that will provide full security where needed. These zones are not affected by the activation of the other two zones and can only be deactivated by switching them off at the control panel. Usually, except in the more expensive types, their role cannot be changed.

The bungalow cannot be properly guarded by the system shown in Figure 4.5, because zone 2 must be omitted to allow the users to go to bed. This leaves a large gap in the defence of the home. To get around this problem, the smaller bedroom and kitchen door switches must be connected to zone 4, with the main bedroom door on its own zone. Although the control box is fitted in the bedroom, you do need to omit or bypass the door switch, otherwise anyone needing to go to the bathroom will have to deactivate the system – no easy feat when you are in the dark, bleary eyed, disorientated, or simply in a hurry!

Don't forget the external alarm box. Wherever it is situated, it should be as high as possible, fitted in a prominent position, where it can be easily seen.

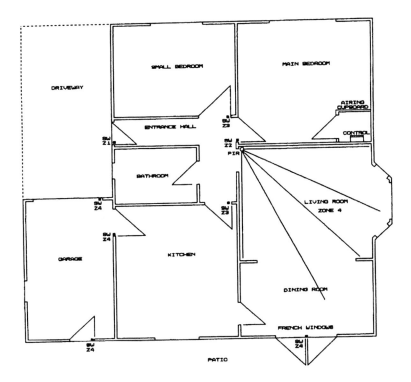

Figure 4.5 Low security bungalow alarm system

Medium security level

The higher levels of security described in Figures 4.6 and 4.7 can be achieved by adding extra sensors and modifying the zone allocation. All downstairs areas are now covered for the detection of internal movement. The sensors will detect anyone entering through a door or window.

When the system is set and the occupier is out, all zones will be activated. Any intrusion, other than through the front door, will immediately sound the alarm. Entry is only allowed to the areas temporarily inhibited by the entry delay. At night, zone 4 is omitted or bypassed to allow movement on the first floor landing and between bedroom areas,

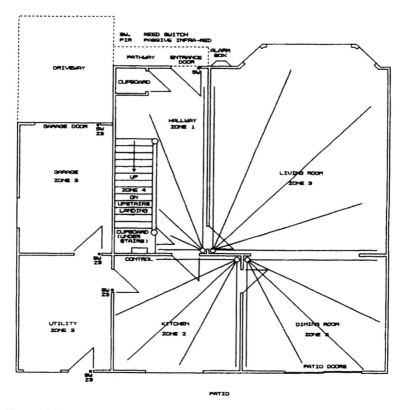

Figure 4.6 Medium security alarm system (house)

leaving all of the downstairs fully protected. Every morning, the occupier must make his or her way to the alarm control box to deactivate the system before entering any other ground floor rooms. The sensor in the hallway will detect the householder and start the entry timer just as if the front door had been opened.

Bungalows are, by comparison, much harder to secure. Accommodating movement around the house with the alarm partially activated can be difficult. The interconnecting corridor between the rooms will usually lead from the front door to all of the other internal rooms. This means that any sensor fitted in that area must be turned off to allow

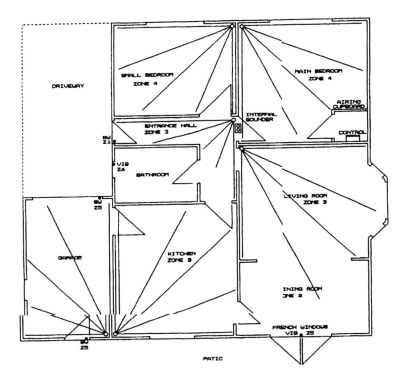

Figure 4.7 Medium security bungalow alarm system

the occupier to move around. The system shown now uses a six-zone control unit, but it is still set and unset from the control box in the main bedroom. At night, the alarm will be set to cover the areas not in use, which would be the bedrooms in zone 4. Now, however, we also have better control over the facility to disconnect any one area. Limited perimeter protection can be established by bypassing or omitting all zones except 1, 5 and 6. This would sound the alarm if any of the outer doors or the bathroom window were disturbed.

Maximum protection

The security plans in Figures 4.8 and 4.9 show how the utility room
and garage areas of both homes have been upgraded by fitting an IR
sensor in each area. Vibration sensors on all ground floor and other 'at
risk' windows create a good perimeter defence. At night, these sensors
and other outer-door switches can be activated to completely seal the
interior of the home. During an alarm call, an internal piezo electric
sounder located centrally in the hallway will generate a very loud wall
of sound that will shock, deafen and disorientate any intruder. Personal

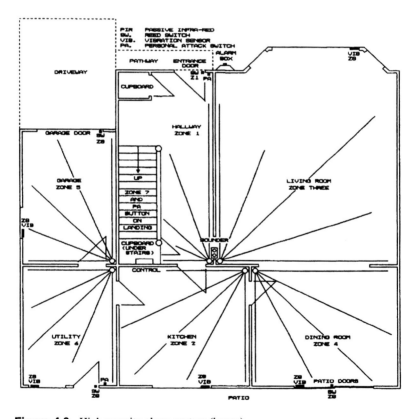

Figure 4.8   High security alarm system (house)

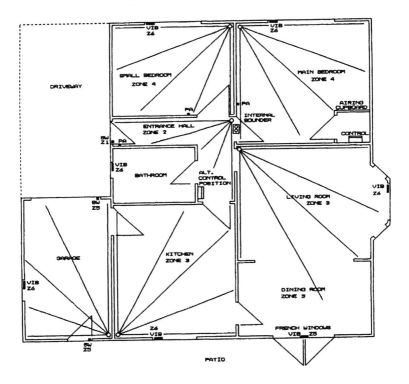

Figure 4.9 High security bungalow alarm system

attack buttons have been placed at strategic positions, such as the front and back doors, the hallway, main bedroom, etc. When they are fitted near to a door, the occupant can press the button if he or she feels threatened by a caller. On the other hand, if someone is lying in bed and they are convinced there is someone attempting a break-in, then pressing the button will sound the alarm.

Where there are more than two floors in the building, the use of a remote control keypad will give the ability to set the system wherever a unit is fitted. If the building is large, they can be fitted near entrance doors, where they can be accessed without having to venture too far into the interior. In this case, walk-through zones can be kept to a minimum and the main control panel can be hidden away in some quiet corner.

Installing the system

Use the information booklets provided with the control unit as a guide for making connections to the panel. Some through-connections, such as those for the tamper circuit and sensors in series, will be made off the circuit board. Each panel type may be different, so take particular note of how open-circuit sensors, such as pressure mats, are wired in.

Initial considerations

It is worth repeating that the control unit should be fitted out of sight, and also in an area other than zone 1, wherever possible. A unit that utilizes remote control pads can be fitted in the loft of a bungalow, where it would be virtually immune from attack.

Think carefully before you start about how you can run the cables to the panel. You should use a cable from each PIR and switch sensor and a single cable looped from each of the window vibration detectors. By using separate runs to as many sensors as possible, the system can be modified at any time in the future.

Many alarm systems have been wired with some of the cables hidden underneath the upstairs carpets. This approach would have saved the original installer's time and money. Unfortunately, the practice puts wiring at risk of being damaged by pressure and abrasion, or from being pierced by carpet tacks. Faults of this nature can easily occur and be hard to trace. To reduce cable problems, it is considered best to run the wiring underneath the floorboards wherever possible. This may seem a little daunting at first, but it does make a much better, neater, more reliable and permanent job in the long run.

If the house is unoccupied, then this will be the best opportunity to carry out all of the wiring work. If, however, the house is occupied, then the carpets will need lifting above ground or lower floor room sensor positions. Furniture will probably require shifting and the cupboard where the control box will be fitted must be emptied. Our bungalow can have another problem, in that many lofts are full of old household items – these will need to be moved around as the wiring progresses. You must also be careful not to step between the joists – you may easily hurt yourself and it can be costly to have the ceiling plastered again!

General wiring practices

When you have decided on which sensors to use and on the positions that afford the maximum coverage, lift the floorboards as close to each position as possible and above the route where the cables are expected to run back to the control box. If a joist needs to be drilled, make the hole as near to its centre as possible – don't make a notch on the top, because this will weaken the joist and may leave cables vulnerable to fixing nails or screws. If there are any mains lighting or power cables nearby, try and leave a 60 mm gap between them and the alarm wires, to avoid any induced interference that could trigger the alarm. When wiring is not installed beneath the floorboards, it should be kept out of sight. Wherever possible, use permanent built-in cupboards for vertical runs from one floor to another. If this is not feasible, then it will be necessary to use white PVC cable trunking. This is likely to be 25 mm × 16 mm and may be bought with a self-adhesive backing and a clip-on lid. After peeling off the protective backing, it will be ready for fitting and, by simply pressing it hard against the wall it will stay in place permanently. This is much quicker, cleaner and just as strong as using wood screws and plastic wall plugs. For horizontal runs, where it is impossible to use the space under the upper floors, trunking can again be used. Sometimes, however, there are other means of hiding the wiring: coving or picture rails can be used as camouflage; if the mess can be tolerated, the cables can be chased into the wall, using a bolster chisel to cut a thin groove along or up the wall, which will be filled in with plaster afterwards. Where all else fails, it is possible to run the occasional cable along the top of a skirting board, although this is not recommended.

Run all of the cables to the sensors first, then down to the control panel position. Any wiring in the loft should be clipped down neatly.

The external alarm box cable may need a hole drilled through the outer wall. Using a 12 mm × 400 mm masonry bit, drill through the wall from inside the loft. Thread a straightened out wire coathanger into the hole and secure the alarm box cable to it as shown in Figure 4.10. If the cable is to be fed down to the alarm box where the roof edge meets the wall, then a different approach will be needed. Drill the soffit board (below the gutter fascia) close to the wall and feed a length of cable up into the hole. Go into the loft and, with a length of heavy wire or piece of wood with a wire hook on the end, 'fish' for the cable. When you have it, tie the alarm cable to it, return outside and pull the cable out of the loft.

Most sensors will be fitted into the corners of a room and very close to the ceiling. Drill a small hole (about 6 mm) up through the ceiling plaster and into the space between the joists. Cut a small length of four-core wire, tie a knot into one end and feed the other end up into the hole. Now when you search for the cables beneath the boards it will be easy to find the sensor positions. Mark each cable to identify its zone or origin. This will save time and a great deal of frustration.

Occasionally, it may be necessary to run wires down the cavity inside

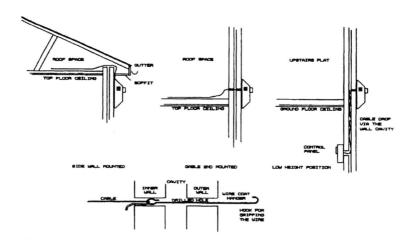

Figure 4.10 Various methods of alarm box wiring

the external walls. Long drops may be difficult if cement from the bricklaying process has oozed out. If a length of string weighted with a heavy piece of metal is lowered down, then it may snag. Should this happen, take out a full brick from the wall in the loft space and, using a flashlight and mirror, check for obstructions. Sometimes, the retrieval of the string or cord from the cavity can be even harder – a half brick taken out from behind where the control panel is to be fitted may help considerably, allowing you to fish for the string with the wire coathanger.

Cable core identification

The wiring examples in Figure 4.11 show how the various sensors can be connected. There are drawings of both single and multiple sensors in series on zone circuits, showing clearly the power, signal and tamper

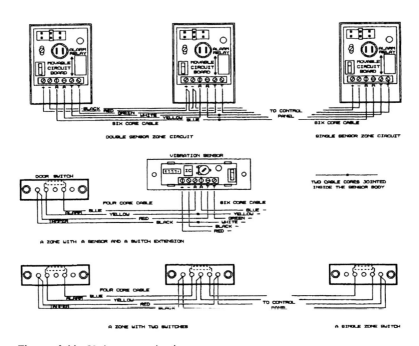

Figure 4.11 Various zone circuits

connections. Colour coding of the cable cores should be standardized throughout the installation. On a six-core cable to active sensors, red and black can be used for the power supply, yellow and blue for the alarm signal pair, leaving green and blue for the tamper loop. For four-core wiring to the switch sensors, blue and yellow could still be used for the alarm pair, but red and black (if used) would revert to use as the tamper loop pair. The cable leading up to the external alarm box would be slightly different again. Wire the tamper loop in yellow and blue, the beacon strobe supply in green and white and the red and black pair to the siren. If a SAB is used, there may be small differences. Consult any supplied leaflets, then whatever colours are used should be recorded on a sheet of paper and kept for future reference.

Fitting switch contacts

Wooden doors and door frames are best suited to the flush-switch type. Drill the hole to take the switch barrel, then offer the switch into the hole and mark around the rectangular outer plate. Use a wood chisel to cut a recess deep enough to accept the plate. This should be cut so that the plate is flush with the frame. Drill a 6 mm hole through the frame from the back of the hole, then feed the cable through and con-nect it as shown in Figure 4.11. The magnetic half of the pair is fitted into the door exactly opposite the switch.

The surface type of switch sensors are usually fitted to plastic or aluminium doors and frames. Orientate the switch and magnet for the neatest fit, connect the cable and screw down the switch half. Keep the gap between them as small as possible with the pair in the same orienta-tion (vertical or horizontal). The cable can now be run back to its point of emergence from the wall, etc. and either held with clips, or, if it is laid on plastic, Superglue can be effective.

PIR fitting

Take off the cover from the sensor, then remove the PCB from the case and put them both in a safe place. Fit the back of the case into position by using wood screws and wallplugs on a brick or block wall. Keep them near to the ceiling to hide the wires. Sometimes, they need to be fitted on a swivelling bracket to allow the sensor to be pointed in the correct direction – make sure the bracket is tight and that the sensor is not free to move. Pass the cable through one of the 'knockout' holes provided, then strip back the outer sheath by about 50 mm and the inner cores by 4 mm. When the wires have been fitted into the connecting strip, fit the PCB back into place and refit the front lens and cover. Be careful not to touch the sensor itself as any contamination could impair the operational sensitivity of the device.

Vibration sensor

The cover of the sensor is removed to reveal a plastic base holding a PCB. Ease out the PCB and place to one side and set the base in the selected position. Fix the base to the window frame. If the window is constructed from PVC or aluminium, then small holes will be needed for self-tapping screws. The connections will be very similar to those on the PIR sensor and you should use the same colours for the same functions. Clip the PCB back into place and replace the cover. A solid foundation is required to allow vibrations to reach the unit without any attenuation.

External alarm box

Mount the siren or bell on to the baseplate using self-tapping screws or nuts and bolts. The tamper switch will need 25 mm × 6BA bolts for mounting on the base. If an SAB unit is used, then two holes of the correct diameter will be needed to mount the spacing pillars provided. Take care not to scratch through the protective wax layer that covers all of the parts.

Although the box will usually be designed to reduce windage, it should be fitted using long screws. If possible, use rust-proofed types – there is nothing worse than an alarm box blown off a wall and left dangling on the end of a cable high in the air.

Internal sounder

Surprisingly, many people have slept on when their alarm system has sounded. They have been blissfully unaware that someone had been trying to enter their home. Equally unbelievable is the fact that, when a siren is sounding at the front of the house, it will seem to be miles away to an intruder attempting a break-in at the back. An internal piezo sounder has a piercing sound that seems to penetrate the walls and windows of a house. The occupant will be awakened from the deepest sleep and a prowler will be made aware that they have been detected. Fit this sounder in a central part of the house, such as the entrance hall or other passageway.

Control panel fitting

Solid fixings are required here to prevent the panel from being hammered or levered from the wall. Use No. 8 woodscrews, about two inches (50 mm) long and plastic wallplugs. If plastic trunking is used to cover the cables, fit this into place first. Feed the cables into the case, tighten down the base then proceed to sort the cables into zone groups. It may be a good idea to connect one zone into the panel at a time. The alarm can then be tested using that zone before going on to the next. Internal panel connections, such as series zone wiring, may be either soldered or joined with a strip of small connectors.

Mains connections

Note: extreme care must be taken when working on mains supplies. If you are in any doubt, enlist the help of a qualified electrician.

The power used by the alarm is minimal and can be taken from several sources. There may be a 13 A power outlet nearby or a handy lighting circuit. The control unit may even be installed next to the consumer unit or fusebox. If the nearest supply is the fusebox, then a 1 mm PVC twin and earth cable must be run from the panel input to a fuse or circuit breaker with a maximum of 6 A rating. The supply mains switch **must** be turned off at the source, i.e. at the consumer unit mains isolating switch, when making **any** connections to the wiring.

To take a supply from a wall socket outlet, you will need to use a fused spur unit. This should not be a switched version, because if it was accidentally turned off, then the battery would run down and the house would not be protected. This spur unit can be mounted close to the control panel, or near to the point of connection, such as a ring main or radial outlet. It can be recessed into the wall using a metal or plastic box, or mounted on the surface. The use of a junction box will enable you to tap into exposed cabling and will save disturbing the existing outlets.

Another method of obtaining a mains supply is to make a direct connection to the lighting circuit. This can be effected by using a junction box, or by taking a cable directly into a ceiling rose. You must be certain that you are not connecting into the cable coming up from the light switch, as this will not give the full supply voltage normally, and no supply at all when the switch is turned on! Figure 4.12 shows how all of these connections can be made. Make sound contacts and remember the use of green and yellow earth sleeving is mandatory to stop any short circuits to either the live or neutral cores.

Do not make this final connection until all of the other work is completed and the alarm is functioning properly.

Wiring outbuildings, etc.

Running a cable to an outbuilding or shed can usually be carried out by digging a trench and laying a length of waterproof tubing. Electrical plastic conduit will do the job well, but garden hose will also do

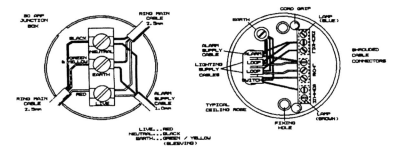

Figure 4.12 Connecting into existing mains circuits

the trick and may cost less. In order to run the cable through the pipe, have someone hold one end while you push the cable down the tube from the other.

Dig the trench to a depth of about 18 inches (500 mm) and lay down the tubing without any twists or unnecessary bends. Place flat tiles, slates or similar on top of the tube to stop a spade from cutting through it. Do not combine the alarm cable with the mains supply to the shed, garage or outhouse, as this will almost certainly cause the alarm to trigger by inducing mains spikes into the wiring.

If the garage or shed is very close to the house, it may be possible to use a short length of rigid conduit, high up. Two holes directly opposite each other, large enough to take the conduit, must be drilled first. Cut the pipe about six inches longer than the gap to be bridged. Fit the pipe into place and, from the house, push the cable through the tube into the garage. Seal the entry into the walls with cement and/or flexible sealant.

If the ground is solid concrete, or the shed is some distance away, a wireless system can be purchased to provide the link. The receiver is connected directly into one of the zones on the control box, but the transmitter is fitted in the building to be protected, with the addition of the sensors to be used.

The sensors in the shed or outbuilding should be wired on the same zone. This will make it easier to install and find any faults. Again, use the wiring details described in Figure 4.11.

Tips and reminders

Keep the cables, even those under the floorboards, neat and tidy. Fill in any holes in the ceiling or walls with either a white plaster-type filler or a flexible sealant. Any gaps between the wall and a length of white PVC cable trunking can be filled in with sealant and smoothed with a wet finger or cloth. Make tight secure connections in all cases. Active sensors and switches must not be loose or prone to vibration.

Read the instructions provided for all the devices used and especially those for the control panel. Learn how each unit can be adjusted to give the best results in the location and conditions in which you have installed them.

Chapter 5

Testing and maintenance

The condition of any electrical system should be checked on a regular basis. The period between tests will vary from years for ordinary household wiring to a recommended minimum of 12 months for an alarm system. Again, the booklets supplied with the equipment should be consulted whenever any work is being carried out.

Testing

It is essential to test a newly completed system. This will reduce the incidence of false alarms or other teething problems. It is presumed that the installation has proceeded well and the circuits have been tested individually as they were connected. Any simple circuit faults have been corrected as they have arisen. Once the last connections have been made and the cover has been placed on the control panel, the alarm system can be tested as a whole.

Sounder and strobe

Most manufacturers have included simple, but effective, checking facilities. Read the information booklet and digest the instructions – this will help reduce any confusion when setting the system to your requirements.

Enter the test programme and select outputs as the sounder and beacon strobe operation. Any add-on circuits should be disabled at this point. It may be an advantage to have a partner or friend on hand to check that the external siren is indeed working correctly and thus help keep the sounding time to a minimum. Indeed, it may not be possible for the person carrying out the tests to hear the external siren while standing at the control panel, because the internal sounder is also ringing.

When all of the sounders have been checked and verified as working, then the beacon strobe operation can be confirmed. The flash rate of the average strobe is around one flash per second. If the rate is much slower than this, then the unit may be faulty or there may be a voltage drop on a long cable run – if this is the case, then the siren may not be working at its best either. There are few ways to correct this: one is to run a new cable from the control panel and double up with the cable cores, thus doubling the capacity and reducing the voltage drop. Another method would be to retrofit an add-on relay driver and power supply as described earlier. The likelihood of this problem occurring is, however, remote and it will only be a problem on very long runs, or where sirens have been used which draw a relatively high load.

Internal panel sounder

Sometimes the control panel internal loudspeaker can be accessed and tested to confirm its correct operation.

Room sensor operation

Depending on the type of control panel in use, the method of testing zone inputs may vary. A built-in programme containing all of the necessary test routines is commonly available on the more sophisticated types, but if the unit is a basic analogue panel, there may be none at all. One certain way of testing any system is to trigger the alarm by deliberately opening a protected door, or moving within range of a space or room sensor when the alarm is set. This method, while effective to some degree, is not really efficient or neighbour friendly. In a test programme, there will most likely be a routine for live walk-testing the sensors. This means exactly what it says: you walk and test the coverage and sensitivity of the sensors room by room. Start by walking across the protected area at the furthest point in the room from the sensor. The test LED should illuminate, and, depending on the panel type, there may be an audible confirmation that the panel had indeed received a signal back from the sensor. Continue to walk slowly back and forth and check that the sensor can pick up your body heat in all of the protected area. Especially check for sensitivity near doorways and windows. When this has been completed to your satisfaction, move on to the next room and carry out the same exercise, and so on.

Door contacts

Enter the test routine as before and have your friend or partner stand near to the door which is to be checked. If a room sensor is in the same area, it will be necessary for that person to either operate the door from the other side or move only the slightest amount sufficient to pull and push the door. If the contact breaks when the door is moved only slightly, then the two parts will need to be repositioned closer to each other. Should this not be possible, then the door latch may need to be tightened or some other method should be employed to stop or reduce movement. Any problem here must be overcome, because any air movement may be sufficient to trigger the alarm.

Vibration sensor sensitivity

Window-mounted sensors can be subjected to a wide range of vibrations. Among these may be heavy traffic, wind driven rain or hailstones, and even ground settlement. Only the vibration caused by knocks and bangs during an attempted break-in should be strong enough to trigger the output circuit. Test the sensor by tapping the window or window sill in various places with the handle of a heavy screwdriver or light pin hammer and observe the results. If the LED switches on with a very light tap the sensitivity may be too high. On the other hand, if a heavy knock is needed, the sensitivity may be too low. Trial and error is needed to reach the best sensitivity. Adjust the trimmer as per the manufacturer's instructions. When you are satisfied that the unit is at the optimum setting, carry out the same procedure at the next sensor.

Personal attack buttons

If the attack circuit had been open circuit when the alarm had been first switched on, the panel could not have been reset. Should the wiring be short circuited anywhere, then one or more of the switches fitted would not operate the alarm. With the panel in the unset or day condition press the panic switch button. The alarm should trigger immediately. Release the mechanism with the key provided, then enter the user code at the nearest keypad or the panel. This should silence the sounders. Reset the system and test the next button for operation. Most panels will not reset until the button has been physically released from its latched condition. The buttons should also work when the alarm is in the set condition (or at any time).

Chime testing

Any system which has the facility to have chimes on the zones should be checked to ensure they operate only on the zones where they are required. The volume may be adjustable and may need to be turned up to enable an occupant to hear clearly from an upstairs room if a door or sensor is disturbed downstairs. Read through the panel instructions and enlist the aid of your helper, who should open and close selected doors or walk through predetermined areas. For maximum effect, the chime should be clearly audible throughout the building. Larger, rambling homes may benefit from having loudspeaker extensions to make certain the chimes are heard.

Default settings

Factory set conditions will apply for all time delays and code settings. To change them, carry out the instructions given. When the system is powered up for the first time with the battery, the alarm will most likely sound. Enter the default code to stop it.

Exit delay timer

The exit delay is of a nominal 30 seconds duration. Where some control systems have a 'final door set' facility, others will not set anyway when a zone switch is open or a room sensor is still detecting the occupier's presence. Because of this last function, it is usually recommended by the manufacturer that the user should listen to the exit delay beeps or buzzer to ensure the alarm has set. Failure to do this could result in the alarm being 'hung' with a door left open somewhere – the alarm would still be unset and the house left open to intrusion. Only when it is known that the system is reliable can the user begin to relax this attention.

Entry delay timer

Many people will be nervous about the time they have available to them for entering the building, entering the codes and disabling the alarm. It makes sense to measure the time it actually takes to enter the house and enter the code. Take care that the slowest person using the alarm is catered for. There is no use setting the delay for the minimum time available when the other family members cannot reach the control panel in the allotted time.

Final checks

It should now be possible to connect and test any add-on circuits and make certain they are operating as expected. Take it easy and try them one at a time. This way, any new problem will be identified immediately.

Maintenance

Recording information

The need for carrying out electrical resistance tests and logging the results may not seem to be apparent at first. Recording the details of each zone and other circuits should not be left to memory. The wisdom and effort of doing all this will be rewarded at a later date when a false alarm occurs and the system needs to be checked for faults.

After the alarm has been checked thoroughly for correct operation and all circuits are completed, arm yourself with a resistance test meter and screwdriver and switch off the mains power. Open up the panel but don't disconnect the battery. Starting with zone 1, disconnect the alarm trigger circuit wires, measure the resistance value and write it down along with the zone number. Continue until the resistance values of all the zones have been recorded. Don't forget to include the tamper and personal attack circuits.

After all of the readings have been taken, fill in a test sheet (like the one provided in Figure 5.1) and log each result. Another copy is provided at the back of the book for you to use (see p. 145). The circuit details of each zone, the types of sensors and where they are situated should all be entered and listed. Any add-on circuits may have their wiring diagrams included, along with the manuals or leaflets of each device used. Don't forget to include the installation date but **never** include details of any codes used.

DATE OF INSTALLATION:

ZONE	AREA PROTECTED	PROTECTION DEVICES USED	TEST VALUE Ω	1st YEAR	2nd YEAR	3rd YEAR
1						
2						
3						
4						
5						
6						
7						
8						
PA						
TAMPER						

Figure 5.1 Installation and maintenance log

Yearly checks are often requested by insurance companies. It is unlikely that they would accept any results submitted by you. If a contractor is called in to do the work, any information given to him will speed up the process and hopefully save you money.

To give you peace of mind you can still carry out the tests and record them each year. For instance, trends may be seen in wiring and connection resistances, which may betray corroded or rusted screws. Any increase or build-up in the values will almost certainly be due to one or the other.

Top of the list is the external alarm box: it is subjected to all the weather can deliver; the cover can be bleached to a lighter shade. A new cover can be fitted in minutes, and a white one cannot bleach.

Dampness in the tamper circuit microswitch can oxidize the contact surfaces and eventually lead to an open circuit. Replace this with a new switch if damage is suspected and change the fixing screws at the same time.

Both the siren and beacon strobe would have been supplied as waterproof and will be unlikely to throw up problems, but check their wiring connections for rusty screws. If the resistance increases here, they will simply stop working due to a lack of supply voltage.

If an SAB has been installed, inspect the wax protective layer for scratches, which will be betrayed by rust or corrosion. If the unit has signs of such damage, clean the area and try to reseal with new wax taken from a candle.

The back plate should have been fixed with rust-protected woodscrews. Mild steel versions will be nearly rusted through after 12 months subjected to the acid rain. Replace them with stainless or brass for a much longer service life.

PIR sensors are often subjected to a daub of paint or two at decorating time. If paint is seen to land on a sensor lens, wipe it off immediately. Don't let it harden, especially if it is a gloss type. Using thinners may cause the lens to lose its ability to transmit IR rays on to the detector. Scratching and scraping the paint away will have the same overall effect. If the lens is damaged in any way, have it replaced. Your local wholesaler will stock the various types for your sensor. Spiders and other insects entering the IR sensor casing can cause false alarms by walking across the detector surface. If there is any reason to suspect infestation, don't spray the inside with insecticide; use a powerful vacuum cleaner, then check again for eggs stuck to the PCB or case. Plug any holes in the enclosure with silicone; this will stop any recurrence.

Whenever badly corroded terminals on door switches are found, replace the switch half then seal it completely with a silicone sealant, as dampness must have been seeping in.

Control panel faults tend to be limited to poor connections. Tighten each screw in turn and check any soldered joints. It has been known for fuses to become slack in their holders and cause faults, loss of battery power or total panel failure. Extract the fuse, pinch the holder grips together and replace the fuse, making certain of a tight fit. Some panels used flat touch pads to control the system – a check may reveal that the most used number keys may be damaged, thus rendering the system useless.

The battery should be checked, but there is really only one test that can be sensibly carried out. Set your voltmeter to DC on the minimum range which includes 12 V. Disconnect a battery terminal and check the voltage. A fully charged example will give a reading of about 13.8 V. Reconnect it and switch off the mains supply and test it again. With a load across the terminals, the voltage should be steady. Should the reading sink slowly down to about 12 V, or drop lower, change it. It is possible that one of the cells has been short circuited or the plates have phosphated; either way, the battery is in need of replacement.

Cable faults should be rare, especially if they were run under the floorboards as suggested. Where the occasional cable had to be run on the surface or beneath a carpet, it may have been damaged in some way. Although a fault would appear eventually, the use of the full alarm system would almost certainly be denied. Where a cable is run beneath a carpet in a doorway, check for abrasion damage to the outer cover, core insulation and bare wires. An open circuit would stop that zone from becoming clear when the alarm is set, whereas a short circuit will stop some or all of the sensors from triggering the alarm.

An area especially sensitive to dampness is the outhouse, shed or garage. Cold, damp, night air will cause accelerated corrosion – check all connections thoroughly. Insects love these spaces and may consider a sensor casing a warm place for a nest. Check for otherwise unnoticed damage caused by vandals or prospective intruders attempting a break-in.

Tips and reminders

If there are any procedures for repair and maintenance recommended by the manufacturers, use that guidance first. Replace any faulty units as soon as possible and, wherever possible, purchase the same type or model as before. Any panel fuses must be replaced with the same rating and type or operating speed. Damage may occur if slow blowing fuses are used instead of fast types.

Any stand-alone devices, such as radio link transmitters or receivers, should be separately checked and tested on a regular basis. Inspect aerials and power supplies and carry out any other recommended measures.

Chapter 6

Existing systems

Older alarm systems can often be recognized quite easily from the outside of a house. A rusty old alarm box conceals a bell that is likely to be corroded and capable of only a feeble warble. Down in the control panel, an ancient circuit board lurks with components oxidizing away quietly. Old, out of tune ultrasonic movement sensors and badly damaged, worn pressure mats are scattered around the house, connected together with worn cables.

This scene is repeated in many homes. The occupants, who know little about electricity or electronic components, are unaware that their alarm is almost useless. The design has probably been out of date for years and will offer only token resistance to an intruder.

Of course this may not be the state of your system. Ask yourself this: Can I rely on my alarm to wake me if my home is intruded upon? If the answer is in any doubt, then you should either install a new security system, or, if the wiring permits, you can upgrade the existing installation.

Upgrading an alarm

The first priority is to check the physical condition of the wiring. If the cables were run under the floorboards, then they will most likely be in good condition. The main contributor to faults in this case are central heating pipes installed after the alarm was fitted. Sometimes, the cables may have been resting on blistering hot pipes causing the insulation to crack, burn or become brittle. This wiring must be in good condition, otherwise the installer must fit a new system from scratch.

Before starting any work, locate the mains supply to the alarm and isolate it. Disconnect the battery to avoid short circuits and, while checking the cable condition, take the time to trace out the wiring routes and find out which sensors are in each line or zone. Write this information down as you go, adding in the number of cores in each cable and possibly the colour coding used.

Many older systems used a switch sensor on each of the internal doors and pressure mats in areas where increased security was desired. Take out all of the old pressure mats, internal door switches and any wiring leading nowhere. If the cables to the door switches are still in good condition, it may be possible to extend the cable up the door frame and position a PIR above the door. The cable could then be cut into the plaster and filled in again to completely hide it.

Switch contacts fitted to the external doors can be replaced with a similar pattern, and any gaps around them can be made up with a filler.

If the system was fitted with early vibration trembler switches, remove these and replace with new electronic versions. The connections to the older types may have been a series loop, using only two of the four cores available. Where there is only one four-core cable leading to each sensor, be suspicious of hidden cable joints. Use the second pair of wires for the power supply to the sensor.

New circuits would consist of any new external doors or those in garages or conservatories. An internal sounder and a couple of personal attack buttons would be welcome additions.

Taking down an old alarm box can be awkward. The easy way to take one down is to lever it from the wall with a crowbar. Don't forget, if the box is made from mild steel it will be heavy, so be prepared for the sudden weight in your hands and don't let it overbalance you. It would be better to let it drop than for you to fall.

Check the number of cores in the box cable. If there are six, then you can use an SAB unit, but if there are only four, then you are restricted to connecting the siren and strobe on one pair of wires, with the second pair used for the tamper switch. It is not a good idea to omit the tamper switch and use separate pairs of wires for the siren and strobe. Alternatively, a second cable can be run.

Strip out the old control panel, taking care not to damage any of the cables leading to it. Throw it and the battery away, together with all the old equipment. You may need to make a trip to the local waste processing centre.

Take care when fitting the new control panel – take out any moulded cut-outs provided and ease the wiring into place. Mark the wall for the new fixings then lift the enclosure clear while drilling the holes. Fit the wall plugs, replace the panel case and screw it back against the wall.

First connect the external alarm box wiring. Connect the battery leads one at a time. If the alarm is sounded, then enter the code to stop it. Add the new internal sounder and panic switch cables and test them. If all is well, connect the zone wiring one at a time, checking each one in turn before trying the next.

Fault finding

That elusive fault that triggers the alarm at every time of the day or night can be extremely annoying. Tracing, isolating and eliminating it can be just as frustrating. Most modern control systems have memory available to recall which zone or circuit triggered the alarm call. Older analogue units had no such facilities.

If a cable break is suspected, then the only way this can be detected is to take resistance readings of each circuit. If one value is substantially greater than the others, keep a record of the reading over a period of a week or two. If the value increases during damp weather then it is possible that there is a badly corroded joint somewhere in the circuit.

An intermittent break that causes an alarm may be vibration dependent. Have your partner or friend walk around, jump up and down on the floor or thump the walls where cables are run. This has its comical aspects, but the real result may be in finding that elusive fault.

Older PIR sensors are much more likely to suffer from scuttling (or nesting) insect syndrome. Any old PIR sensors should be cleaned out thoroughly.

Extending and adding zones

As long as the wiring runs are suitable, then extra zones can be added easily. The actual extension wiring involved does depend on the number of cores already at the first sensor. Figure 6.1 shows how another zone can use the same cable. A four-core cable can carry two switch contact zones and a six-core can be used for three zones. As many switches as necessary can be connected in series on each zone to complete the circuits. Simply loop in the wires as shown here, and previously in Chapter four. For extra information, read all the preceding chapters.

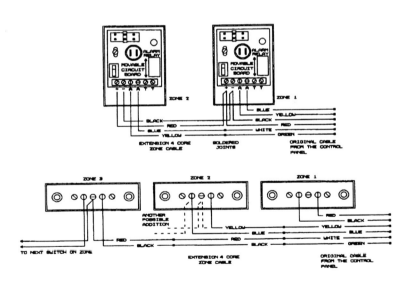

Figure 6.1 Adding zones to original wiring

Chapter 7

Security lighting

Illumination of the garden area or drive can be welcoming to home owners returning from work or a night out. If it is an automated floodlight that switches on as you approach, then it would be equally startling or off-putting to any prowler or prospective intruder. The usefulness of exterior lighting should not be underestimated especially when a properly installed system could be used to drive internal sounders to warn you of a prowler, or operate an external siren to call a more general alarm.

The same floodlighting can also be used to show the way on dark winter nights; they can light up an icy path and make the going safer. Using these floodlights can be easy: they can be operated manually or automatically; be switched on permanently or for a predetermined period. Those who live on farms or isolated homes can use the bright light to work under, or create an oasis in the darkness during the long winter nights.

Correct positioning and adjustment is vital for correct operation. Think of how often sensor-driven floodlights switch on as you drive or walk past a house. This isn't funny; because the lights are *not* being switched on by someone on the property, the validity of the lighting is lessened – everyone expects them to switch on at any time for any reason, and not for security. Think also of the electricity bills.

It should not be forgotten that there are, in many gardens, an abundance of items worth stealing in and around the garden, patio or back yard: furniture, pets, garden ornaments, hanging baskets, prize vegetables, clothing etc. It should be remembered that some items can be very expensive to replace – clothes for example, can run to hundreds of pounds worth, and they are simply left hanging on a line. Ornamental fish can have astronomical values, yet they are left swimming around in a hole in the ground. This is where external lighting systems can give real security to the area around the house. Several circuits will be described in this chapter. All are useful in their own right, and, depending on the situation, can increase the security of any home.

There are other exterior sensor-operated luminaires or light fittings available, such as enclosed coachlamps, open spot or flood, screw in lamps and the more utilitarian bulkhead types. These tend to have a rather limited light output and, often, the sensor is fixed. The beam shape may be wide or narrow, depending on the type of lamp (flood or spot) or it may give a diffuse spill of light. They are really only decorative and suitable for illuminating the immediate surrounding area.

The halogen floodlight

Providing instant brilliant light with a reasonably natural colour rendering, easily controlled, and available exactly when it is needed; the halogen lamp comes in various sizes or ratings from 150 W to 1500 W. The lamp itself is a coiled tungsten wire set into a linear, halogen gas-filled, quartz glass tube fitted with ceramic connectors at either end.

These lamps do not need any starting gear and give full brilliance immediately. The sketch in Figure 7.1 gives the general construction details of a typical luminaire. The enclosure is usually a zinc or aluminium diecast body with fins to help dissipate the intense heat generated. A specially shaped dimpled aluminium reflector directs the light in an angled rectangular beam towards the target area.

A hinged frame supporting a glass panel covers the reflector. The primary purpose of the cover is to keep dampness out.

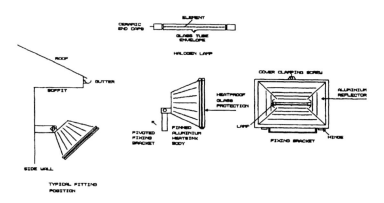

Figure 7.1 Typical enclosed halogen luminaire

A pivoting stand-off bracket is supplied to mount the unit onto virtually any surface. Care must be taken when mounting the lamp to meet certain installation criteria. The information supplied with the lamp will almost certainly insist that, when installing the lamp high up on a wall, it must not be fitted closer than about 12 inches (300 mm) beneath any wooden structure, such as the soffit board. This is to prevent ignition or slow burning of the wood due to the intense heat rising from the lamp. The lamp must also be level, or within a few degrees of horizontal. This increases the life of the lamp by stopping the element from stretching at one end and bunching up at the other.

Changing or replacing an expired halogen lamp is somewhat trickier than normal. If possible, choose a dry calm day and set your ladders up against the wall at the same height as and to the left of the light. (If you are left-handed then set them up at the right of the lamp.)

Turn off the power to the system at either the mains or controlling switched spur and, if possible, remove the circuit fuse. Unscrew the fixing bolt on the glass fronted lid and allow the cover to hang downwards on its hinges. Using a sheet of kitchen tissue paper, grip the lamp tube and push to one side against spring pressure. When the lamp end is clear of the holder pull it out and away from its normal fitted position.

Fitting the new lamp is a reversal of the above procedure, but keep your fingers well away from the glass. In many cases a piece of foam or a protective paper cover is wrapped around the tube so that it can be handled immediately. If no protective cover is supplied, use part of a clean sheet of tissue.

Touching the glass of the bulb can leave minute quantities of grease on the surface which will burn the tube away slightly and seriously weaken it or cause a fracture. The glass can be cleaned using a tissue dampened with methylated spirit or similar fast-drying cleaner.

The mains supply

The minimum requirements for wiring the supply up to the lamp with manual operation is a fused supply and a means of disconnection. If a rewireable, or cartridge, fuse is used then it should have a rating of no more than 5 A. Where an MCB (miniature circuit breaker) is connected into the circuit, the rating should be 6 A maximum.

The cables used for any of these lighting circuits that carry mains power must be the PVC-insulated type of house wiring cables with copper cores, having a minimum cross sectional area of 1 mm and an integral earth wire.

The means of disconnection may be a standard lighting switch, or it could be the switch on a fused spur where the supply is taken directly from the ring main. It may be possible to use the normal lighting circuit to supply the floodlamp, but in some cases this may not be suitable. The services of a qualified electrician will be needed if there is any doubt – he or she can test the cabling and verify that it is satisfactory and safe.

The external PIR sensor

There are a few crucial differences between these and the internal types – waterproofing, increased load-switching capacity, access to variable user controls and, of course, mains operation for most types. Figure 7.2 shows how a typical sensor looks and how it can be fitted to an external wall.

Protection against dampness is essential to the longevity of the device. Corrosion can creep around the sensor PCB causing open and short circuits. In time, this may show up as erratic operation or as an intermittent, fuse-blowing fault.

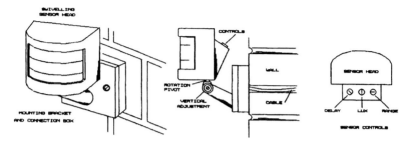

Figure 7.2 External passive infra-red sensor

Increased load-carrying capacity is needed to directly switch on the floodlamp. To do this, a heavier duty relay is fitted into the sensor. It must be noted that the contacts are normally rated to supply one 500 W lamp or about 2 A. An extra relay circuit would be required if there is a need to drive more than one lamp from a sensor.

The controls for adjusting the sensor's parameters are fitted for external adjustment to make it easy to get access and to save the weatherproofing seals from being disturbed. These will normally include the timer for setting the 'on' period of the lamp after the sensor has been triggered, and 'lux' level control for setting how low the ambient light must be before the unit will detect movement and switch on the floodlamp. There is sometimes also an adjuster that allows the user to set the 'sens', or sensitivity, of the detector to reduce any false alarms due to wind, or moving trees or animals. The lux level control can be set to allow the sensor to operate even during broad daylight. The usefulness of this is twofold: it enables the installer or user to test the coverage and also to use the device for detecting visitors and triggering sounders at any time. Both of these facilities will be discussed later.

Sensor coverage

The adjustment bracket allows the sensor to be pointed where maximum sensitivity is needed. Some types have a full 120 degrees of coverage with a volumetric lens; others have the same scope, but boast a blanking-off facility, where the sensing 'beams' can be tailored by the use of plastic IR 'shades'. Specialist sensors can have tight beam patterns similar to those of the corridor lens, which can be set to detect movement in a very specific position. Your car could be protected in this way – any prowler or car thief moving around the parked vehicle will be detected and illuminated.

Sometimes it may be desirable to set the sensor in such a manner that it will detect only movement close to a house wall: this is known as the curtain mode. This is useful where the legitimate movement of your neighbours could trigger the floodlamp unnecessarily and generate false alarms. The protection of a window, or the illumination of a narrow passageway between houses, is likely to need this type of setting. An ordinary sensor can be used if the top row of lenses are used while the others are blanked off with tape.

Separate sensors and floodlight

This arrangement will give a large degree of flexibility in the positioning of the two units. The sensor can be situated where someone can be best detected, with the least chance of them avoiding the device. Similarly, the floodlight can be in a location that will afford the best all-round or localized illumination. One example might be the driveway of our example house where the sensor must detect the presence of a person or vehicle entering the gates. The lighting should not be in such a position that will blind the driver or impair his or her view, but will provide a brightly lit path.

If installed in a patio or rear garden, the lighting should be user friendly, especially if the light is switched on automatically when the occupier takes a few steps from the patio door. This can be achieved with the sensor 'looking' across the door area. As with internal sensors, the best sensitivity is gained when the beams are crossed by a moving heat source. The illumination of a patio would be in the area which would provide the best spread of light consistent with good security.

The wiring circuit involved is shown in Figure 7.3. The supply is taken directly to the sensor. Another cable is then run to the flood from the switched output of the sensor. As with the manually operated

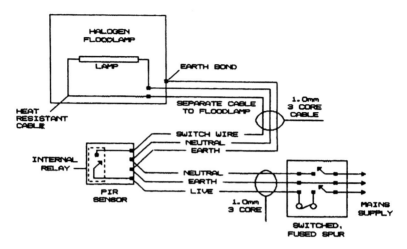

Figure 7.3 Minimum, basic wiring to separate sensor and flood

floodlight the power must originate from an isolating switch with a fuse not greater than 5 A rating.

Combined commercial units

These ready-assembled units can be found in many DIY or hardware shops. They are easy to fit, needing only a satisfactory supply and a setting to suit the location. Cheaper units, including the single sensors, should be avoided if possible – they only offer a limited lifespan. The more expensive types will cost up to twice that of a cheaper combined unit, but will last for much longer and be more effective.

The wiring in Figure 7.4 shows how the combined unit can be supplied. This is essentially the same installation as for the single floodlight, the only difference being that, once switched on and set up, the power is, in normal circumstances, left on permanently.

There can be limitations when installing combined units. The most likely one is simply that, because the sensor and floodlight are built as one unit, it may be very possible that one or the other is not in the best

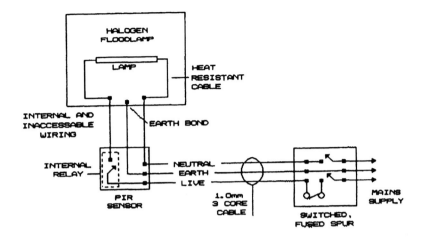

Figure 7.4 Combined unit internal wiring and supplies

position for detection or illumination coverage. This problem must be considered before installation, or preferably purchase.

Sometimes combined units can be switched on manually by flicking the control switch. This might involve switching the switch off then on again within a short period, such as two or three seconds. This will energize the relay, forcing it to remain energized until the switch is again turned off and on to return operation to normal.

It should be noted that, when the sensor positions are being determined, there may be a cause of future problems nearby: heat emanating from the central heating flue can trigger the sensor.

Illumination coverage

It should be remembered that the beam from the floodlamp will spread outwards to create, in an ideal situation, a very large rectangular pool of light. The area on the ground that is illuminated by the brighter part of the beam is known as the footprint. The spread of light and the size of the footprint will vary according to the shape of the lamp reflector and the fixing height.. The higher the lamp, the larger the area covered.

Conversely, the higher the lamp the lower the brilliance at ground level (see Figure 7.5).

One way to test the span of light is to wire up the floodlamp with a length of cable and plug it into a convenient upstairs socket outlet. The lamp can then be held out through an open window and the coverage noted. Try and work out whether the lamp will need to be higher or lower, or moved to one side or the other to get the best results.

Planning the lighting and sensor positions

A long look at the house plan in Figure 7.6 will reveal a number of locations for separate sensors and floodlamps, and for combined units.

We can start with the combined unit installed above the bay window. The front garden area is well covered by both the sensor and the lamp with one exception: the narrow space immediately beside the bay has a shadow area around it. The significance of this depends on the area, other natural cover and any general street lighting. The width of this shadow will be reduced as the sensor light is raised. It may be possible to reduce it further if a fixing bracket can be made or obtained to hold the unit away from the wall. This will, of course, make it harder to reach.

The sensor half of the separate pair above the drive should be sensitive to most movement starting from just inside the gate and up to the

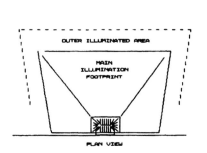

Figure 7.5 Illumination footprint area

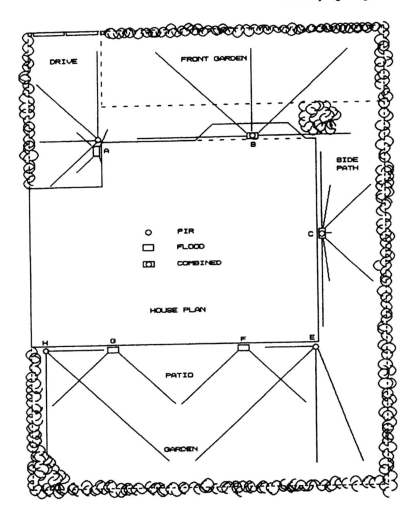

Figure 7.6 Plan view of lighting and sensor coverage

corner of the house. This could overlap the area covered by the garden PIR above the bay window, giving continuous protection across the whole front of the house. The floodlight is angled almost directly downwards to the drive in front of the garage to give the maximum amount of illumination just where it is needed. Because of this positioning, a driver will not be blinded by direct light. I have seen these lamps fitted on garages facing down the drive and completely blinding the driver as he or she approaches the garage – this is both unnecessary and dangerous.

The combined unit located at the side of the house has both the sensor and lamp pointing downwards. This helps to secure the side of the building by using the sensor in a curtain mode and reduces light overspill into the neighbour's property. It should be noted that light overspill from poorly angled and positioned lamps can affect relationships with neighbours. No one likes to have their bedroom or living room suddenly lit up when all they want is privacy.

The two lamps and sensors sited above the patio area of our house should give the area total protective coverage. The spaces covered by the sensors overlap in the centre of the patio and also cover from the wall and doors up to the back hedge. The illuminated area should, if the lamps are correctly positioned along the wall (and also for height and angle), only reach up to the boundary hedges. These can be controlled independently or together, depending on the control circuits used.

Customized control circuits

The circuits outlined here can enhance the basic concept of the sensor-operated floodlamp. Three simple but very useful methods of control are sketched out in Figure 7.7. The first drawing shows how a simple mechanical or electronic time switch can be connected into the circuit, together with a commercial photocell. This combination allows the lamp to be switched on when daylight is reduced but only within the time frame permitted by the timer. A manual switch bypasses the timer to give extra control. There are times and places where this arrangement may be useful, such as on remote or isolated homes or farm buildings, where the farmer may need to work long hours on dark evenings. The lights could even be programmed to switch on and be ready for an early morning start.

To set this system up, decide on the maximum operating hours you would want the lighting to be operative for on an automatic basis. This could be from 4 p.m. to midnight, then again from 5 a.m. until about

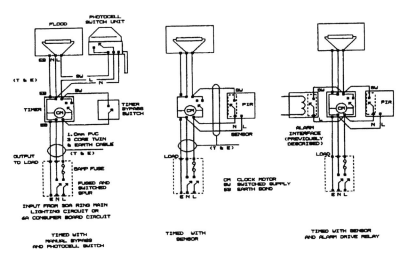

Figure 7.7 Various security floodlighting circuits including photocell switched, timed, sensor and alarm driven

8 a.m. These times should cover both winter and summer conditions. The photocell will then switch on the light when daylight starts to wane, but the clock timer will in turn switch them back off again later.

A simple extension of this idea is to couple a sensor into the wiring to create a semi-automatic system that operates outside the timed hours it was programmed for. This could partially overcome the problem of varying daylight hours over the year by bypassing the time switch.

There is another very useful addition to these two circuits, which would involve connecting an alarm interface relay as described in Chapter Three. This will only switch on the light if the alarm is activated, and then only when it is dark if a photocell is fitted, or alternatively it could be connected in such a way as to override both the time switch and the PIR sensor.

Manual control can be added in by connecting a standard single pole switch across the timer switch terminals, together with the other devices. If there is too little space or overcrowding of cables in the general area of the timer, then a four-way lighting junction box can be employed to reduce congestion, which is bad practice and a source of overheating and burnt-off connections.

In Figure 7.8 warning sounders are added into the control circuits to tell us when there is anyone in sensing distance of the external detectors as it is possible to not notice one of the outside floodlights switch on.

The easiest way to add in a sounder is to use a relay with a coil wound for mains operation. This would be connected between the switch wire (leading to the lamp) and the neutral return. The buzzer circuit would be connected to a normally open pair of contacts. Thus, whenever the lamp is switched on, the sounder will operate. This will continue until the integral sensor timer switches off the lamp again. For simplicity, the buzzer may be powered by a battery. The whole unit may be fitted into a plastic double socket box set into the wall. A plastic cover with a buzzer control switch and slots for the sound would finish off the job.

Another, but uncommon, method of adding a buzzer into a circuit (and it is the only way to retrofit it into the circuit of a combined sensor light unit) is to construct a simple current-operated relay.

A large-pattern open circuit reed switch can be used for this. About twenty to thirty turns of 18 SWG enamelled wire are wound directly onto the glass body of the switch. This carries the current of one 500 W lamp sufficient to generate a magnetic field strong enough to activate

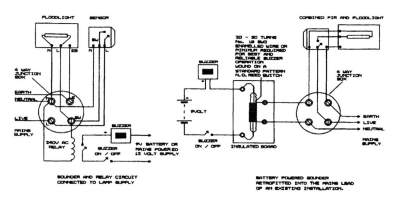

Figure 7.8 Two methods of connecting low power internal sounders to a sensor-operated floodlight

the reed contacts and sound the buzzer. Do not overlap the wire and keep the ends of the windings at least half an inch (12 mm) from the reed switch ends. Again, the circuit shows a battery-driven sounder. This should be simple to install, but adding a power pack could be easy if a cheap battery eliminator is bought.

Breaking into the wiring and making connections can be managed by using a junction box as shown. Use a plastic box to house the relay and, above all, keep to recognized methods of construction.

Another sketch in Figure 7.9 describes how to connect and wire a floodlight into a circuit controlled by a timer and PIR with manual switching of a light and buzzer. This gives full control of all of the functions from one convenient position. A standard three gang lighting switch controls all of the functions. These select timed/manual, PIR on–off lighting and buzzer on–off. There are no practical instructions given

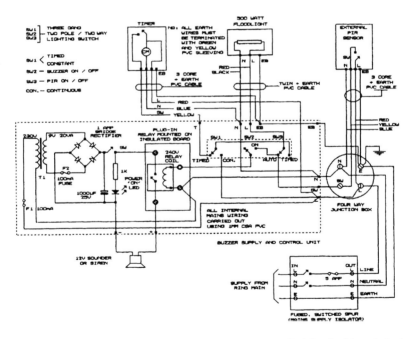

Figure 7.9 Wiring and circuit diagram of sensor-controlled flood with switched sounder

for constructing the unit. Again, this is left up to the individual, but the circuit and the associated wiring are clearly shown. The sounder can be an internal piezo sounder or a high power external electronic siren. This would give a prowler a bit of a shock if he or she came into your garden and triggered the PIR. The sounder can operate independently of the lamp and work at any time of day or night if the lux control on the PIR is adjusted properly. This could then be used for 24-hour protection of the garden areas. Another useful modification would be to use a two-way switch to select between internal sounder or external siren as the need dictates.

The buzzer or siren supply is a basic 12 V DC PSU permanently driven by the mains via the switched spur. An LED shows the PSU status and confirms that it is ready and active. This unit could be mounted out of the way in the loft space and the wiring for the switch control run down to the desired location.

To connect two separate lamps and two PIR sensors together so that either of the sensors will switch on both lamps simultaneously, and to have the facility of manual separate switching, the addition of a load boosting relay will be needed. This circuit, which is ideal for our patio area, is shown in Figure 7.10. The relay is connected to the outputs of

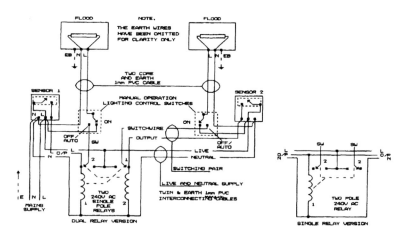

Figure 7.10 A two-lamp, two-sensor system with common, PIR operation

both sensors. The two lamps are connected to this common supply line via a two-way changeover switch. This ensures that both lamps will normally operate together, but if the changeover switches are operated, independently. When either of the sensors detect a moving body and energize their internal relay, the output supply will drive the external common relay. The contact rating is then increased from a nominal 2 A for a 500 W load to a minimum of 1000 W for a 5 A relay. If more than two lamps are needed, then the relay capacity could be increased accordingly. The cable size may need to be increased if the runs are long.

Most sensors use a relay to switch on the lamps. A few may use a triac or other solid state device. If your sensor does not click when the lamp is illuminated, then it may contain one of these. In this case, two relays will be needed to isolate the two devices and operate the above circuit. The two sets of contacts should then be connected in parallel.

Testing

Walk-testing the completed lighting systems should be possible at almost any time of day or night if the sensors have lux controls fitted. The action is similar to that of checking the internal alarm PIR sensors, except that the floodlamp or other luminaire will turn itself on, giving a positive visual indication of its operation.

Turn the time delay or 'on' control down to the minimum setting then start by walking around the perimeter of the protected area. If the lamp switches on at almost every step when you are moving across the sensor view, then the sensitivity will be satisfactory. If it is possible to move out of the area beyond a fence or wall, try walking as before and check the lamp remains unlit. This should establish that the sensor will not pick up any movement outside of your property. Any 'dead' or insensitive spaces should be noted and an attempt to orientate the sensor to correct this should be tried.

When the sensor coverage has been fully checked and adjusted, it should now be okay to adjust the lamp 'on' period. Four to five minutes is an average time delay. Any shorter that this may cause the lamp life to be reduced: any longer than this (and the sensors often come with at least a 10-minute delay) and your electric bills will start to escalate.

Check that any sounders fitted to the system operate properly and that they switch on or off as you would expect. Confirm that the switching controls also operate as they should, then label them with their function. Labelling avoids confusion later.

Finally, ensure that everything is watertight. Grease and silicone sealants will be helpful. Remember, any ingress of water can cause a leakage current, which could prove fatal in some circumstances.

Summary

Controlling the lighting is by four main methods:

- a manual switch to override the system at any time;
- a clock-driven switch to limit the operation to certain times of the day;
- use of a PIR sensor to detect prowlers;
- a photocell switch to give that extra automatic switching when darkness falls.

Chapter **8**

Video security and door entry systems

Video security cameras can be installed in and around our homes. They can tell us who is at the door and, if necessary, record events to tell us just who or what it was that was prowling around in our garden when we were out, why the sensor-operated lights switched themselves on and the sounders.

For those who fear to answer doors, or who have little faith in door chains, a remote audio/video system would allow them to give access only on positive confirmation of a visitor's identification

Cameras and their control

Television has come a long way in the last 60 odd years. For most of this time, we relied on cameras that generated an analogue picture signal. The Vidicon camera was, for many years, the favourite in the security world. It used a photoelectric tube with a scanning electron beam. The beam was modulated by the light and dark variations of the image on a photoelectric plate. This signal was amplified and transmitted along wires to a monitor that could convert this video signal directly into a picture.

Today's equivalent is smaller, much more sensitive, and due, to the electronic circuitry built into it, uses far less power. The CCD, or Charge Coupled Device, camera is a solid state instrument that electronically scans a silicon light-sensitive cell. The cell is a thin wafer etched into rows and columns of pixels (picture elements) that make up the complete image. There are many advantages over the analogue camera:

* low power consumption;
* precise picture geometry;
* higher sensitivity;
* low susceptibility to magnetic fields;
* resistance to image burning.

This latter phenomena occurs in cathode ray tubes after very long periods of showing the same image.

The various CCD sensor sizes give varying numbers of pixels and picture lines, depending on the manufacturer. Generally, the larger the CCD size, the greater the picture resolution.

The output from a monochrome camera is a composite of video and internally generated synchronizing signals and can be taken from a coaxial connector. These signals drive the monitor directly with screen control information, such as the number of lines per frame, etc. Alternatively, the camera can be operated by an external synchronizing signal source, if more than one is used, to create steady images when changing from one camera to another.

The power supply is either 9, 12 or 24 V DC, or 240 V AC. Power requirements should be decided on when planning the system. The 240 V AC version can be plugged in almost anywhere, but the DC types are safer, due to the lower voltages involved. A minimum of cabling in this case would be a coaxial video cable and a DC supply twin. The sketch in Figure 8.1 describes the construction of a typical camera and its main component parts.

Miniature cameras

Some of the smaller cameras are simply a CCD image sensor mounted on to a PCB with a basic wide-angle lens for focusing, together with

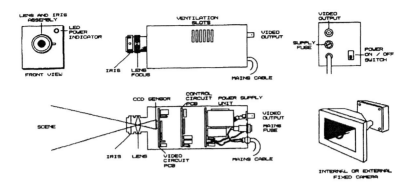

Figure 8.1 Typical CCD video camera

the associated electronic circuitry. They can be fitted into very small enclosures and installed in such a manner that they are not readily noticeable. At least one such camera has been fitted into the same enclosure as a PIR sensor; this effectively combines the two systems, and events can be recorded if the PIR detects a prowler and the alarm is sounded.

Lens types

The function of the lens is to place a focused image on to the sensor. There are various types with differing basic functions, such as magnification, angle size, etc.

Among the many lens types are fixed, varifocal and zoom. With the basic lens, there is a need to specify the focal length to suit the view. A focal length of about 4 mm gives a wide angle image. This is suitable for covering a large room or an outside area. For an image that gives a view similar to that of the human eye, a 6 mm version is needed. An 8 mm lens will give a narrow, long-range picture, just right for corridors or close ups. The angle of view will usually be stated, along with the other lens information.

Getting the right focal length for any specific scene need not be difficult. The angular field of view of a 4 mm lens (wide angle) is about

99 degrees. At around 10 m distance, the camera view will cover a little less than 24 m across. The following simple formula can be used to determine the horizontal field of view, to enable the correct lens to be purchased:

$$f = 2 \, (d \tan [\Theta/2])$$

Where:

d = distance from the object or scene (m)

f = horizontal field of view (m)

Θ = angular field of view (degrees)

Figure 8.2 shows the differences in scene coverage using the three focal lengths given above.

The manually adjusted varifocal lens is very popular and is adjusted initially to focus the image and then left fixed in that position. This allows the camera to be installed and adjusted to suit a diversity of applications.

When specifying a lens, it should be noted that they are available in sizes that suit the CCD sensor size. This means that a half-inch lens can have different focal lengths, but is intended for a half-inch imager.

Iris

Another consideration is the iris size. This allows a specific amount of light to reach the sensor. Modern cameras have sensitivities that start

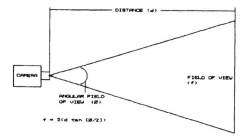
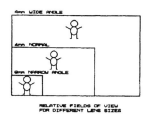

Figure 8.2 Determining the viewed scene width

in fractions of lux (or light level). Too much light will overload the sensor, giving an over-exposed image; too little and there will be a featureless picture.

The manual iris shutter, or stop, can vary the admitted light level. It must be set to match the lighting at the chosen location. This is only suitable, however, when the light conditions are static. Frequently, the iris is an automatic feature. An electronic iris is self-adjusting and has the ability to cope with varying levels of natural or artificial light. This should keep the image contrast at a constant level.

Zoom lens

A powered lens and iris assembly with a zoom facility can be obtained. With a magnification factor of eight or more, the scene can be viewed in close up or wide angle. A typical zoom lens with an 8:1 ratio can change the effective focal length from 6 mm (wide angle) to about 48 mm (close up, narrow angle). The adjustment is carried out using a miniature DC motor to drive the lens assembly via a remote controller.

Siting the camera

The best position for a camera is in the highest security risk area, e.g. the back of the house, the driveway, the front door, or other frequently used entrance, and perhaps even certain rooms in the home. Figure 8.3 shows another plan of the ubiquitous house, but this time with suggested camera positions.

Camera position 1, situated near the front door, covers part of the drive and the doorway. Position 2, near the garage door, will give a full view of the driveway, to the gate and into the road beyond. The patio area and the two back doors are covered by the camera in location 3. An internal view from camera 4 covers both the dining room and lounge.

Other locations inside the home could include a nursery room. This would be much more effective than a simple baby alarm. If the occupier is self employed and needs extra security in a home office, or there is an expensive computer in a games room, then a camera could give the added 'insurance'.

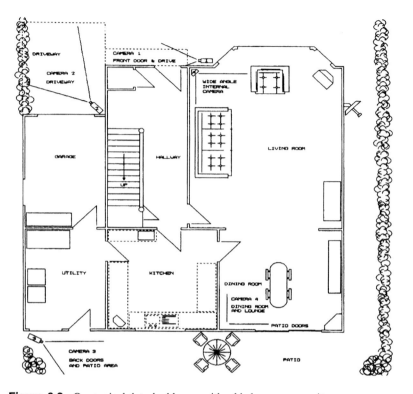

Figure 8.3 Our typical detached house with added camera security

If a camera is located close to a halogen lamp, then be careful not to install it above the lamp where the heat may cause damage. Overloading the CCD sensor with direct light will also have a devastating effect. Place the camera to one side of the lamp, where neither the heat nor the light can do any damage.

Coverage

The coverage of a camera can be likened to that of the floodlight footprint. An added advantage of setting up and adjusting a camera yourself is the ability to see in real time exactly the scene you want.

To do this, you can connect the camera video output directly to a monitor that has been set up temporarily. A mains extension cable and a length of coaxial cable will be needed. The image can then be set for focus and coverage as you view. This is much better than trying to rely on the judgement of an assistant, who is attempting to relay a description of the scene from a remote location.

External camera protection

Figure 8.4 shows a typical external camera fitted inside a protective enclosure. Weatherproofing is best provided by a commercially made sealed aluminium unit with a plastic window with shading provided by an extension of the top cover. A wall mounting bracket allows the camera to be swivelled vertically and panned horizontally, then locked into place.

The more modern external camera types are becoming much more compact. The case is likely to be sealed and weatherproofed and should not need any extra protection. The plastic viewing window may be made

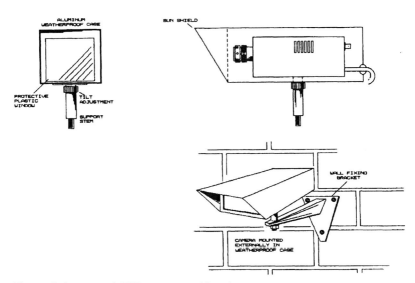

Figure 8.4 External CCD cameras and housings

from material with IR filtering properties. Figure 8.5 shows various models of both external and internal cameras in this range.

The choice of camera must be made to suit the situation. Inconspicuousness may well be desirable. Dark cases will go well outside, whereas white or cream-coloured types will suit interior locations. Any camouflaging should reduce the likelihood of the camera being noticed and becoming a target for vandalism.

Sound

Some cameras may have a microphone incorporated in the system. If the camera is installed outside, then it could be used as part of a door-answering system. The same idea could possibly be used as a very sophisticated baby alarm.

If there is an attempted break-in and the alarm system has triggered a video into recording mode, then the voices of the villains will be heard when the tape is reviewed, which may give extra evidence and could lead to an earlier arrest.

Video switchers and camera control units

To enable the user to view more than one scene without changing signal cables, the switcher was developed. Integrated circuits used as electronic

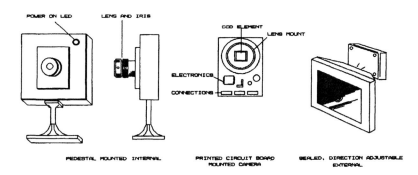

Figure 8.5 Modern internal and external cameras

switches increase reliability and flexibility. In Figure 8.6 such a switcher and its application is shown.

The image from each camera can be viewed in a manual mode by pressing the appropriate channel button or sequentially in an automatic mode by pressing the appropriate 'scan' button.

The central feature is the multiplexer, or switcher, to which, in this example, four cameras, a controller and two monitors are connected. Between each camera and the switcher are the interconnecting coaxial video signal cables. Between the switcher and the monitor is another coaxial. To control the channel switching,a multicore cable is connected between the controller and the multiplexer.

The multicore carries digital information to drive the electronic switches. This may be in a four-bit binary code format. The individual code can be selected manually, or can be automatically sequenced by a clocked counter. The image scanning rate can be increased or decreased by changing the clock speed. This is normally adjusted by means of a rotating knob on the control panel.

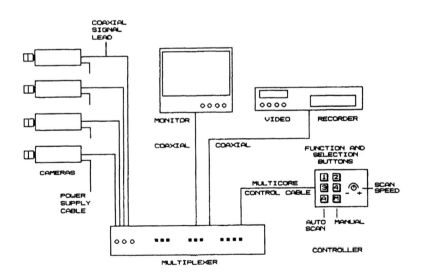

Figure 8.6 Typical multiplexed video security camera system

Video movement alarm

A steady, motionless, picture is boring to watch. There is a device which can 'read' an image and call an alarm condition if there is a change in the picture content. This can be used to sound a low level local sounder or even a full alarm. This attachment should only be fitted to internal cameras because of the false alarms that would be generated by, for example, debris blown around by high winds in an external setting.

Accessories control

Digital telemetry is another medium for transmitting control commands. This is mainly used on professional systems.

Turning the camera in the horizontal plane to scan the scene is known as panning. This can give nearly 360 degrees of view. By raising or lowering the camera angle the operator is able to see over a range of distances. Remote controlled motors are used with a local power supply and digital signals for the movement commands.

To keep the camera window clean, a wash and wipe unit can be mounted, operated from the remote controller. This ensures a clear image in all weathers. A miniature, motor driven, wiper blade performs the action. A refillable bottle containing several litres of cleaning fluid, with an integral pump and control unit are also required.

Video recording

A standard domestic video recorder records images on to a magnetic tape in a series of angled strips, to save tape length. Time lapse recorders have a different approach, as a single frame is recorded every few seconds instead of about 50 per second for the domestic variety. The time savings allows the recorder to use one tape over several hours. A standard E180 three-hour tape is thus capable of a 24-hour recording.

Other facilities offered by manufacturers include:

- selectable recording speed (seconds per frame);
- on-screen menus for setting up the system;

- time display recording;
- movement alarm.

Monitors

Construction

The monitors used for video surveillance and general security are specialized units designed only for this purpose. Most have only basic controls such as brightness, contrast, frame and line hold, etc. They may be monochrome or colour. Sizes range from a 9 inch to 17 inch tube. The case will probably be of a sturdy metal construction for professional use, while the domestic types will most likely be plastic. Some of the specially made home systems come with switchers and other camera control equipment in the same unit.

Inputs

The input signal needed to drive the monitor will have to meet certain specifications and must comprise the picture information plus vertical and horizontal synchronizing elements to create the image. Essentially, the scanning beam of the monitor 'freewheels' until the synchronizing signal is received, then locks into a steady state. Horizontal scanning is at a frequency of about 340–400 lines per frame, although the actual number is determined by the camera output.

The image information is then used to drive modulation circuits, which increase or decrease the electron beam strength and therefore the brightness. It is this change in brightness that creates the monochrome image. For colour monitors, the PAL standard is used in a similar manner to televisions. This gives a picture line resolution of the full 625 lines. Most of the professional security monitors will have a video output socket provided, to enable the signal to be passed on to another viewing station.

Sound

Sound or audio inputs may be provided separately, though some models may use the video cable to carry the signal. This will most likely be a multicore, with the video, audio and even control wires in the same cable. A loudspeaker set into the case then relays any sound picked up by the camera microphone.

Integrated systems

There are many systems for domestic use now on the market that have integrated some of the features described above. The use of modulators to generate UHF signals that can be fed directly into television aerial sockets can replace the monitor. This allows the user to switch between the television channels and the camera signal. Sometimes the AUX or VCR 2 channel may be usable; otherwise an unused TV channel will do.

One system uses a central converter, or modulator, which can combine up to four camera images for split-screen viewing of several locations simultaneously. The advantages of such a feature are obvious.

A recent addition to the market uses on-screen programming and automatic picture break-in with different modes for at home or away use. Some new cameras can illuminate an image with active IR lighting; there are also thermal imaging and low-light level cameras.

Another interesting camera feature is the use of a PIR sensor to trigger the control box. It may be used to automatically change the television channel to the camera signal, or to switch on the video recorder, depending on whether the occupants are at home or away. In this case, the time and date are recorded on screen, along with any sounds.

All operation is by remote control, using stored signals. Remote control codes for other devices, such as the hi-fi or interior lighting, can be programmed, allowing them to be operated automatically.

It should be remembered that these systems may use dedicated devices that are not always readily available. Though they will be robust and have a long service life, future availability of replacement parts may be uncertain.

Door entry systems

Door intercom

In Figure 8.7 there is a very simple telephone circuit, including a description of the two principle components: a carbon microphone and a balanced armature receiver.

The microphone consists of a container loosely filled with carbon granules. Carbon changes its electrical resistance under pressure – the lower the pressure the higher the resistance.

One end of the chamber holding the carbon is subjected to varying pressure from a diaphragm, which moves very slightly whenever sound waves press against it. The resulting change in electrical resistance mirrors the sound impinging on the diaphragm. If a voltage is applied through the chamber, the resultant current flow would vary with this change of resistance.

To reproduce sounds, we need to use a device that will change the electrical variations into mechanical movement. A moving coil loudspeaker would not respond with any great success due to its low

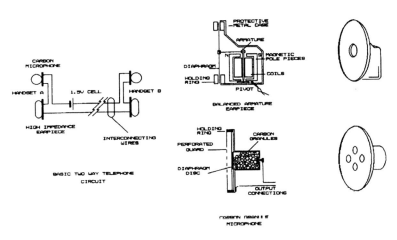

Figure 8.7 Simple telephone circuit and the construction of the magnetic earpiece and carbon microphone

impedance and relative insensitivity. A balanced armature receiver is ideally suited to this use, as very small currents can produce movement and therefore sound.

To use this circuit for a door intercom, quite a few changes are necessary. The visitor may not want to lift a receiver to his or her ear. It is much better to simply press a button, wait for a reply, talk into a concealed microphone and have a voice talk back to you.

In Figure 8.8, there is a circuit for an intercom that can be installed in almost any home. A visitor presses PB1 to attract attention. Inside the house, a buzzer sounds. By picking up the handset, the occupier switches on power to a microphone and earpiece circuit. This circuit relays the voice of the visitor from the door to the internal receiver. Power is also applied to another part of the circuit that amplifies the signal from, in this case, a magnetic microphone and drives a loudspeaker at the door intercom unit.

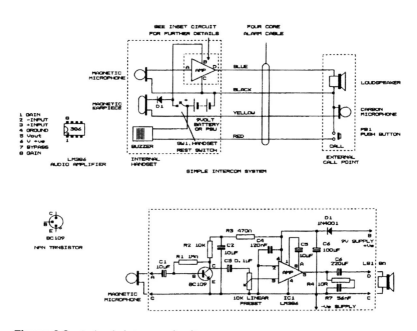

Figure 8.8 A simple intercom circuit

A small IC amplifier and transistor pre-amplifier amplifies the tiny output voltages of a magnetic microphone (which could easily be a high impedance earpiece used in reverse). Amplification increases the signal sufficiently to drive the LM386 IC, which is in turn configured to give a gain of about 200. The preset resistor is included to set the loudspeaker volume.

The internal handset arrangement can be built using an old telephone. The external panel, however, should be constructed with waterproofing in mind. It should also be set into the wall for physical protection. The interconnecting wiring may be a single run of four-core alarm cable or screened type if available.

Commercial systems

The simplest types available have an intercom similar to that described above. A power supply and magnetic door release is included to allow the occupier to open the door from a remote position.

Electromagnetic release mechanisms are fitted into door frames to replace the strike plate of the standard mechanical cylinder lock. Figure 8.9 describes the layout of a typical system. A mains supply is fed directly to a power unit, which in turn delivers a low voltage supply to the internal handset. A cable is run between the handset and the external call unit and carries the intercom audio signals and the call button return. Power to release the door is often taken via a separate cable directly to the release magnet.

The multiple units seen at the doors of blocks of flats are simply a duplication of the single system. Specially made control units have holes already punched out to accommodate the necessary switches and incorporate a single loudspeaker.

Video intercoms

It is now possible to install an intercom unit that is combined with a microphone, CCD camera and loudspeaker for full conversation. At the interior console there is a built-in microphone, door release button

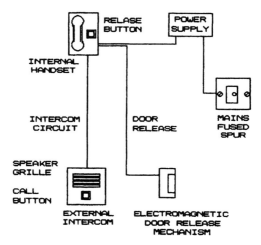

Figure 8.9 Basic door entry intercom system and door release mechanism

and miniature monitor. Although only a few inches in size and in monochrome format, it relays the image of your visitor with clarity. The whole unit is no greater in size than most telephone receivers.

One system only needs a two-core cable to transfer all signals: video, sound, and release circuits. At night, the scene is illuminated by infrared LEDs mounted beside the camera lens. This eliminates the need for traditional lighting and the subject may not know that they are being scrutinized.

Chapter 9

Miscellaneous security gadgets

Various security gadgets and add-on devices are readily available at many local DIY stores or electrical shops. These can include basic plug-in lighting timers, photocell-switched external lighting and miniature, self-contained, stand-alone alarms.

The latter can be either fixed or portable and can enhance the main home system or cover one area that needs a little extra protection. One situation that could benefit from these devices is student or single-person accommodation. Such simple alarms offer protection that would not otherwise be practical.

Self-contained alarms

Magnetic switching door or window alarm

Among the self-contained alarms on the market, these are the simplest in design. They consist of a piezo element and its associated circuit, a timer, and a magnetic switch and magnet. The latter two parts are fitted and operated in a manner similar to the door switches described in Chapter One. The two parts are fixed to the frame and door as before.

An ordinary slide or key switch can enable the device. Any subsequent opening of the door will break the magnetic contact and sound the alarm. The siren will stop after a short period of time and reset itself. Other uses are on sliding patio doors, opening windows, internal doors, etc.

In Figure 9.1 there is a circuit diagram for two very simple low power door or window alarms. The first circuit uses a CD40106 hex Schmitt trigger to generate a pulsing tone and drive a single piezo transducer. The switch is a standard normally closed reed that is held closed by its companion magnet. When the magnet is removed, the pulse circuit is enabled. The output of N1, which is connected in astable oscillator mode, rises and falls at a frequency of about 1 Hz. This, in turn, switches on and off the second astable N2, which is operating at a frequency of about 3 KHz. This signal is buffered by N3 and ultimately drives a 3 KHz piezo transducer. The current consumption should be in the region of only a few microamps, which would make a standard 9 V battery last several months.

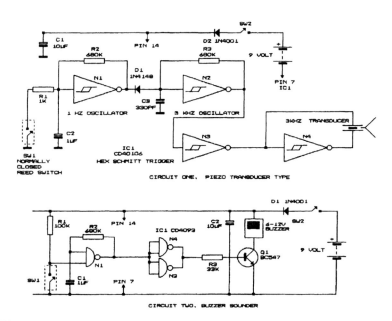

Figure 9.1 Two door or window alarm circuits

In the second circuit is a CD4093 quad nand Schmitt trigger, which is used to generate a single 1 Hz pulse. This is then buffered by two of the remaining gates, which are used to switch on and off a BC547 NPN transistor. The buzzer output is pulsed by the transistor in a similar manner to the above circuit. In both circuits, SW2 may be a key switch mounted on the case for more secure control.

All unused gates should have their inputs grounded or connected to the positive supply line and their outputs left open.

Vibration sensitive alarm

This device operates on the same principle as the vibration detectors described in Chapter One. The alarm is enabled with a switch on the enclosure, and may have a 'settling down' period of a few seconds. Any vibration, such as a burglar attempting a break-in, will trigger the alarm immediately. As before, the siren will time out and reset. These alarms can be fitted almost anywhere in the house.

Equipment removal alarm

This unit will detect when a computer, television or video recorder has been disconnected from the mains supply and moved. The two conditions for sounding the alarm are the lack of mains input and the vibration involved in moving the equipment. Should the mains to the building fail, then the alarm will not sound. Similarly, if the computer or video is knocked or moved without power loss, the siren will not sound.

Timed IR alarm

This device is an IR sensor combined with a timer and siren driver circuits to create a self-contained volumetric room alarm. It is wall mounted for convenience with the sensor 'looking' out into the room. A normal exit timer is built in and is enabled when the device is switched on, giving adequate time to leave the area. On returning, the sensor detects body heat and, after a few seconds will sound the alarm unless it is turned off in time.

Remote control IR alarm

A variation on the timed IR alarm, this unit is switched on and off using a remote control fob: it must therefore be armed after leaving and disarmed before re-entering the protected area. The main advantage of this type is that anyone breaking into the area will trigger the alarm instantly. There is no delay in which the burglar can grab anything of worth. The alarm can be fitted out of reach – even on the ceiling!

Personal attack alarm

There are many versions of these alarms on the market. If they are to have the desired effect, they must be loud. Pressing a button should give a sound level of at least 120 dB, but preferably 130 dB. Out in the open, sirens do not have the same apparent loudness as when they are operated in enclosed spaces. These devices should be kept in a pocket or handbag. It must be remembered that 130 dB is a lot of noise and can damage hearing – direct it away from your own ears.

Lighting controls

Time switches

These timers were briefly mentioned in Chapter Seven. There are motor-driven mechanisms with cams or tappets operating microswitches. There are also electronic versions that can be programmed using a digital LCD. The correct type for any given situation will depend on the application, and particularly whether the lighting is a permanent or temporary installation.

Electro-mechanical motor-driven timers

These are used to switch on and off lamps at various times, to give the prospective burglar an impression that the house is occupied. Normally, the timer is plugged directly into a handy 13 A outlet. The lamp to be controlled is in turn plugged into the timer.

Small pegs or cams are either inserted or lifted to form actuators that press against the lever of a small microswitch. The switch then passes the supply through to the lamp. Each peg or cam will switch on the lamp for a specific period. While most of these timers are for 24-hour operation, there are others on the market that have seven-day dials.

The 24-hour version is commonly incremented in 10-minute segments. In this case the minimum recommended 'on' period is normally for two segments, or 20 minutes. The seven-day timers can be expected to have one-hour increments with a minimum of two hours of operation. The daily timer will repeat itself every 24 hours, but the seven-day version vary for each 24-hour period. Figure 9.2 shows these timer setting methods.

Most mechanical timers run on a mains-powered AC motor and have reasonably accurate time keeping. A big disadvantage is that the motor will stop if the supply is cut. This will put the switching times out of step. Battery backup saves resetting the clock after a power cut or blown fuse. A quartz crystal-controlled low voltage mechanism is powered for up to 100 hours from an internal rechargeable battery.

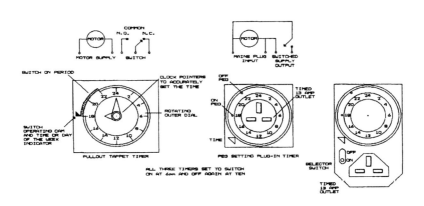

Figure 9.2 Three types of mechanical, motor-driven timers

Other mechanical timers

Cheaper, simpler timers may have only two 'on and off' periods available on a 24-hour dial. This will be satisfactory when used only occasionally, but the timing may be noticed if used for a longer period.

Electronic timers

An electronic timer can give a programme of events that can change daily. This helps to reduce the threat of a prowler noticing the house is unoccupied. The readout is likely to be an LCD designed to give all of the information that is needed to set up the programming.

An internal digital clock will have either mains-derived frequency control, or a quartz crystal oscillator for extra accuracy. The time can usually be displayed in either a 12- or 24-hour format.

Most will offer a repeated daily setting for weekdays if desired. The weekend settings can be different to reflect that change of activities. Some types have the feature of random pattern switching, which makes it seem as if there are people moving around the house.

Some timers are surface mounted and used on specific lights, while others are direct replacements for the lighting switches used for the hall or living room, etc. If programmed with some imagination, it will really make the house seem occupied.

In contrast to the mechanical timers, the electronic variety can be set to the nearest minute. The smallest timed period could then be one minute, although this would be somewhat pointless.

Photocell controllers

This is a common method of controlling exterior security lighting. Public highway lighting is another common use for these, and they can be seen on many street lamps. Some control individual lamps, while others may be used to control a whole street. This practice can be extended for use where homes have large gardens or areas of land surrounding the house. Guide lamps on a long drive could be switched on automatically at dusk.

The standard pattern industrial photocell switch is intended for external installation. It can be used for controlling relatively heavy currents of up 10 A for a resistive load equivalent to about five 500 W halogen lamps – quite a lot of lighting power. An example circuit was given in Chapter Seven.

The switch is mounted under a translucent plastic dome to allow an all-round inflow of light. The dome is also designed to be self cleaning to some extent. The base is usually made from an opaque black plastic with a fixing bracket moulded on to it. Figure 9.3 shows a typical photocell switch.

Light falling on a large area silicon cell allows the cell to conduct sufficient current to power up a small heater. The heater element is wound on to a sensitive bi-metal strip, which bends when heated. The bending action, depending on the light level, opens or closes a pair of contacts, as the ambient light level wanes at dusk, the photocell circuit switches on the lamp. At dawn, the increasing daylight switches off the lamp automatically.

Electronic photoswitch

Similar in shape and function to the above type, this version uses an electronic circuit to perform the switching action. A small silicon photocell is used to drive a relay directly from a built-in electronic circuit. The parameters can be more closely set during manufacture, with very predictable results. The current rating, however, may be lower than the standard bi-metal version, and should be checked before purchase.

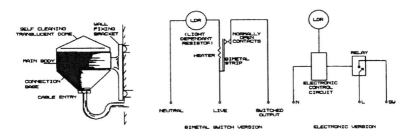

Figure 9.3 Standard-type photocell switch

Automatic luminaires

External light fittings may have built-in electronic photocell control elements. These lights are best used for automatic all-night illumination. The lamp power is generally kept to about 100 W for these units, which will give a large enough pool of light for most situations.

Built into all photocell switches is a delay period of a few minutes. This delay reduces switching caused by dark clouds passing over the sun, or from the headlights of passing cars.

Locations

To avoid incorrect operation, such as early switching on at sunset or late switching off at sunrise, the individual sensor must not be fitted in a position where it will be in shadow at these times.

Wherever possible, luminaires fitted with an integral photocell should be installed with the sensor facing towards the light whenever it is situated in a passageway or other gloomy area.

Chapter **10**

Tools and equipment

The correct tools for any job will not only save time, but can give a better result and a neater finish. Used properly, tools kept in good condition will reduce accidents. The following is not an exhaustive catalogue, but will help to accomplish most of the work generally involved.

Power tools

- **Mains-powered drill.** Choose one with a reasonably high power rating (500–600 W), hammer action and, if possible, a two-speed gearbox. When drilling in wood, disable the hammer action and use the machine at higher speeds.
- **Rechargeable cordless drill.** The use of a battery-powered drilling machine will give portability and freedom and is, to some extent, safer. Unfortunately, the power and speeds available will not match the mains-powered machine. Use this tool for drilling outside, or for small holes in woodwork, etc.
- **Jigsaw.** Taking up a floorboard may be easier by cutting across the board with a jigsaw. You will need to drill access holes for starting the cut. Take care to determine the positions of any water pipes

before starting. Don't use this tool if there is any chance there will be electric cables nearby. Use the correct blade for cutting wood. This will be a large-toothed type, which does not clog with saw dust.

- **Circular saw.** A dangerous tool if not used properly, this can be used to lift floorboards and will do it quicker than the jigsaw. There is a guard fitted over the blade – make sure it is in place and do not take it off. Set the blade cutting depth so that it matches the thickness of the boards, and run the saw along the length of the boards and then cut across the width where you need to lift it. Any nails should be hammered through the board into the joist with a long punch before attempting to lift it clear. Chipboard can be cut into directly. Any unsupported chipboard edges or floor board ends must be held in place with lengths of wood when refitting them. Figure 10.1 shows two examples of how this is managed.
- **Inspection handlamp.** A handlamp is especially useful in loft spaces. A hook fitted to the top of the cage allows the lamp to be hung from a convenient nail. Alternatively a spring-powered toothed clamp can provide grip.

Power tool safety

Care of mains-operated power tools must be taken to avoid risk of injury. Check for damage to the mains lead – an exposed conductor can give

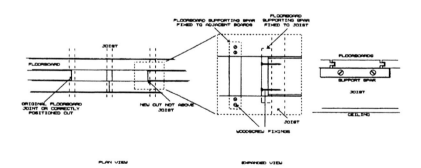

Figure 10.1 Two alternative methods of holding an unsupported floorboard

a fatal shock if touched. Keep the ventilation airways of a power tool clear of accumulated dirt and dust – this will reduce overheating and help prolong the working life of the machine. Do not use mains-powered tools in the rain – a slight build up of surface water can conduct sufficient current to give a fatal shock.

There are three ways of reducing the chances of electric shock:

- use a 110 V tool and matching safety transformer
- use a 240 V tool and an isolating safety transformer
- plug your equipment into the mains via a residual current circuit breaker (RCCB) safety trip that will cut off the supply in the event of an earth fault.

Power tool accessories

- **Flat wood drills.** These are intended for use with high speed drills and should be reserved for cutting rough holes through floors or joists. Be careful though when the drill is about to break through – because the bit is flat, the point will emerge through the partially uncut hole and the cutting blade may grip. This grip can be very strong and the machine can be pulled out of your hands and start rotating. For this reason, do not use the 'hold' button on the drill.
- **Masonry drills.** Use to drill holes in brick, stone or concrete. For the smaller fixings, the following list is a guide:

Screw size	No. 6	No. 8	No. 10
Wallplug colour	yellow	red	brown
Drill size	5.0 mm	5.5 mm	7.0 mm

To drill through brick walls, a minimum shank of 300 mm will be needed or 400 mm for drilling at an angle. The diameter should be no less than 10 mm, although three-core cables will need about 12.7 mm.

- **Long drills.** Standard drill bits are rather short. Long series drill bits are available, but are rather expensive. If a welder, blacksmith or similar is known, then a drill can be welded or brazed to a length of round steel bar. A 600 mm bit with a diameter of about 6 mm will be very useful for cable holes above room sensors.

Hand tools

- **Combination pliers.** A 200 mm pair will cut through all of the cables met. An insulated pair is also useful. The pipe grip can be used to grip nuts or bolt heads while using a small spanner for final tightening.
- **Side cutters.** Generally smaller than pliers, but can strip insulation from thinner wires much more efficiently. A small pair of cutters is very useful for cropping excess wire from the smaller components. They can also extend into awkward places.
- **Knife.** A sharp knife can be used for cable stripping, trimming, etc. Always cut away from yourself. Beware of slicing into cable insulation.
- **Large flat-blade screwdriver.** Also useful as a lever, or hole enlarger in plaster walls or ceilings.
- **Small flat-blade screwdriver.** For general use. A long version can reach into panels or other awkward places.
- **Terminal screwdriver.** An invaluable tool for tightening the smaller connections found in sensors and control panels. A diameter of 3.2 mm with a shaft length of 100 mm is a useful size.
- **Large cross-blade screwdriver.** It can be safer to use crosscut screws, as their design reduces the likelihood of the screwdriver slipping. The drive power should be more positive, as long as the correct size and type is used.
- **Small cross-blade screwdriver.** Some sensors use small crosscut screws for fixing the PCBs, as will other components.
- **Tin shears.** These are invaluable as a cutting tool for plastic trunking and cable access points in the control panel. Used as a pair of scissors, they can shape the trunking body, cut out slots in the side wall and trim the lid to size. They will leave a straight clean edge.
- **Lump hammer.** A two pound (1 kg) hammer is a useful tool to aid in lifting floorboards, cutting holes in walls, or even heavy wood chiselling.
- **Claw hammer.** Ideal for lifting floorboards, pulling out unwanted nails, etc., as well as for general use.

- **Pin hammer.** This is used for tapping home the pins used in cable clips, or occasionally for helping start small screws into wood.
- **Feather cutter chisel.** When taking up a floorboard, the feather, or tongue, between the boards will have to be cut. This wide, thin-bladed chisel will fit in between two boards and will cut through the tongue. It can also be used as a lever. This is a noisy exercise, so use earplugs.
- **Cold chisel.** Only used for cutting into brick or stone, etc. If the house is in the process of being built, then cut holes in the brickwork in the ceiling space on the ground floor, where it would be otherwise impossible to pass cables from one room to another. Also used for chasing cable into walls.
- **Wood chisel.** Use for cutting recesses and grooves in wood for cables, trunking or switches. Keep it sharp for a neat cut.
- **Crosscut saw.** General purpose timber saw.
- **Large hacksaw.** Steel, aluminium and other metals can be cut with this tool. Slotted screws with a rusted or painted head can be cleaned out by running the blade through the groove.
- **Small hacksaw.** Can be used for cutting metals. Plastic trunking can be cut easily, but there will be a rough finish. Use for cutting through wood or metal in awkward and inaccessible places.
- **Crowbar.** This tool can be employed in lifting floorboards and for general heavy duty levering jobs.

Test equipment

- **Multimeter.** For reading the voltages of mains supplies, an AC voltmeter must be used. A moving coil voltmeter calibrated to read within a few per cent of the correct value is an ideal choice. It should cover AC ranges up to at least 250 V, have DC ranges up to the same value, and read resistances at least as small as 1 Ω without difficulty. The scale should be as wide as possible to give a good resolution where it is needed most.
- **Simple voltage/polarity tester.** To check that a 12 V supply has the correct polarity, a simple low voltage LED indicator is used. Using a dual LED with a common cathode connection lead, or two

separate LEDs, a tester can be constructed using the circuit described in Figure 10.2. The first (red) will inform you of the presence of a voltage from a minimum of about 2 V. The second (green) LED will start to light up at about 12.5 V, which is the nominal voltage of the battery in your alarm system. When the voltage applied to the Zener reaches 12 V it starts to conduct. The voltage across the 1K resistor connected to the Zener anode will start to increase. When the voltage reaches about 0.6 V the PNP transistor switches on and drives the green LED.

- A battery may be added to allow for circuit testing. It can be used to verify that circuits are indeed closed, but it must be stressed that a low resistance in the circuit will not affect the LED brightness, but it may confirm the opening or closing of door switches, or sensor relays, etc. When testing for continuity, only the red LED will glow to confirm a circuit.

Miscellaneous items

- **Ladders.** A sturdy pair of ladders, extendable to at least 21 feet (7 m) are essential to reach the height needed for alarm box or floodlight installation. The angle at which ladders may be safely used is judged by a simple formula. The height at the top of the ladder, where it rests against the wall, should be four times the distance from the wall. If the ladder angle is too near the vertical,the

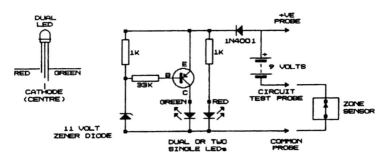

Figure 10.2 Simple DC low voltage tester

ladder may fall backwards away from the wall. Should the distance from the wall to the foot of the ladder be too great, there is a danger the whole thing may slide away at the base. Figure 10.3 shows the safe angle required.

- **Stepladders.** For in-house access to the sensors and sounders. Many modern types have a top ledge for holding tools and equipment, and are multi-positional.
- **Small trowel.** Filling in external holes with cement is best carried out with a small trowel. For internal repairs use it for patching or filling chases in plasterwork.
- **Sealant applicator gun.** Smaller holes and joints can be filled with a flexible silicone sealant. The colour can be chosen to blend with the decor. White is the most common indoor colour, with black or brown for external uses. The filler is squeezed out of a tube via a long tapering nozzle by the ratcheting action of the applicator trigger. It can be smoothed down by using a damp cloth.
- **Soldering iron.** The connections inside the control panel can be soldered to ensure a trouble-free joint. Similarly, those joints inside sensors, where extensions are connected, can also benefit. The standard mains-powered iron is very easy to use, but it does suffer from one problem – a lead. A portable solution is a gas-heated iron. There is only one drawback, and that is the slight risk of igniting leaking gas from other sources.
- **Flashlight.** Use it in the loft spaces for short periods or visits. Combine this with a small mirror and you can easily see along the bays between joists when searching for a cable.
- **Small wire hook.** Fashioned from a wire coat-hanger, this is invaluable for reaching into tight spaces and retrieving cables. Simply straighten out the wire as best as possible, and, using, pliers, form a small loop at one end and a hook at the other.
- **Long reach hook.** Very similar to the wire hook, this is made by taping a wire hook onto a long piece of wood about 20 mm square. This is ideal for retrieving the cable from the far reaches of loft or floor spaces.
- **Safety glasses.** These should always be worn when using power tools or a hammer and chisel on brick, etc.

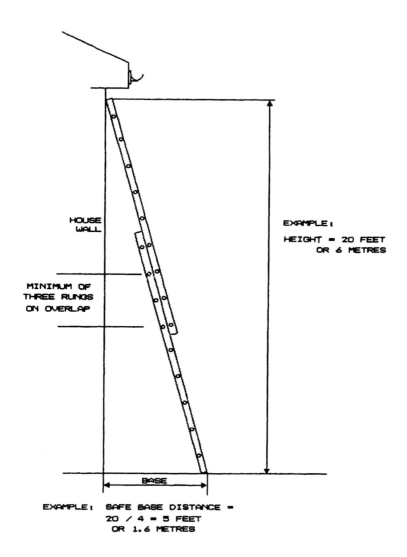

HOUSE
WALL

EXAMPLE :

HEIGHT = 20 FEET
OR 6 METRES

MINIMUM OF
THREE RUNGS
ON OVERLAP

BASE

EXAMPLE : SAFE BASE DISTANCE =
20 / 4 = 5 FEET
OR 1.6 METRES

Figure 10.3 A safe ladder angle

- **Dust face mask.** A build-up of dust or fibre glass over a period of time can lead to chest complaints in the future, so use a new mask every day when creating dust or working in a confined space where there is fibreglass or other fibrous loft insulation.

● **Clothing.** Use a pair of good cotton overalls – they have pockets galore for holding tools and can reduce minor skin abrasions. Wear boots or shoes with toe protection. Your head may need the protection of a hard helmet. Always use approved products.

● **Information.** Read any booklets or paperwork that are supplied with equipment and tools. Take note of any special instructions and adhere to them. If the equipment is hired and there is any doubt about the best way to use the tool, then ask the assistant for advice.

Appendix

Project component lists

All resistors are rated at 0.5 W, 5 per cent carbon or metal film. The non-electrolytic capacitors are ceramic plate 50 V working for all values less than 0.1 µF. Above this value, the capacitors are all 100 V mylar or polyester film. Electrolytic capacitors are as stated in the lists.

Chapter Two: alarm circuits

Figure 2.2. Simple analogue alarm circuit

R1	1K	C1	100 μF 16 V
R2	33K	C2	1 μF 63 V
R3	33K	C3	33 μF 16 V
R4	1K	C4	0.1 μF 100 V POLY
R5	680K	C5	1 μF 63 V
R6	33K	C6	1000 μF 25 V
R7	220R	C7	1 μF 63 V
R8	33K	C8	33 μF 16 V
R9	1K	Q1	BC547
R10	33K	Q2	BC547
R11	1K	Q3	TIP31A
R12	680K	Q4	BC547
R13	220R	D1	IN4148
R14	220R	D2	1N4148
R15	33K	D3	1N4148
R16	33K	LED1	amber 5 mm LED
R17	680K	LED2	amber 5 mm LED
R18	33K	LED3	red 5 mm LED
S1	Single pole key switch		

Suitable enclosure and circuit board to fit

Power supply unit

Transformer	2 x 15 V 20 VA	D4 & D5	1N4001
F1	2 A fuse & holder	ZD1	10 V 400 mW Zener diode
F2	1 A fuse & holder	R19	1K
C9	1000 μF 25 V	LED1	red 5 mm
C10	10 μF 16 V	IC4	7812 (1 A) regulator

Figure 2.3. Add-on circuits

R20	1K	D6	1N4148
R21	33K	D7	1N4148
R22	220R	D8	1N4148
R23	33K	D9	1N4148
R24	1K	D10	1N4148
C11	1 μF 16 V	D11	1N4148
Q4	BC547	LED 4	amber 5 mm LED
Q5	BC547	LED 5	red 5 mm LED
		LED 6	green 5 mm LED

PCB circuit board to suit

Chapter Three: output circuits

Figure 3.3. Sounder circuits

Discrete components single tone sounder

R1	560R	C1	0.022 µF
R2	7K5	C2	0.022 µF
R3	7K5	Q1	BC549
R4	560R	Q2	BC549
D1	1N4001	X1	3 kHz resonator

Low power single tone sounder

R1	56K	C1	3.3 nF
R2	6K8	IC1	CD4011
PR1	22K	X1	3 kHz resonator

High power variable tone sounder

R1	680K	C1	1 µF 63 V
R2	100K	C2	100 µF 25 V
R3	56K	C3	3.3 nF
R4	1K	IC1	CD4011
R5	6K8	Q1	TIP121
PR1	22K	ZD1	9 V1 400 mW
D1	1N4001	TR1/2	loudspeaker transformers
X1/X2	3KHz transducers		

Figure 3.5. Experimental siren circuits

Adjustable depth, sweep frequency generator

R1	1K	C1	3.3 µF
R2	680K	C2	0.1 µF
R3	68K	C3	10 µF 16 V
R4	6K8	C4	10 nF
R5	10K	D1	1N4148
R6	220K	IC1	1NE555
R7	220R	IC2	NE555
PR1	100K linear		

Pulsed tone generator

R1	1K	C1	1 µF 63 V
R2	680K	C2	0.1 µF
R3	33K	C3	10 nF
R4	10K	Q1	BC547
R5	220K	IC1	NE555
R6	220R	IC2	NE555

Two-tone warble generator

R1	1K	C1	0.33 µF
R2	220K	C2	0.1 µF
R3	1K	C3	0.01 µF
R4	220K	C4	100 µF 16 V
R5	5K6	C5	0.1 µF
R6	220R	D1	1N4001
R7	220R	D2	1N4001
PR1	10K	ZD1	9 V1 400 mW
Q1	TIP31A	IC1	NE555
LS1	Siren or loudhailer speaker	IC2	NE555

Figure 3.8. Alarm interfacing to add-on equipment

Additional siren option

D1	1N4001	RL1	12 V DC 2 pole 2 way
D2	1N4001	Transformer	9 V AC 1 A 20 VA
D3	1N4001	F1	1 A and fuseholder
D4	1N4001	F2	1 A and fuseholder
D5	1N4001	F3	1 A and fuseholder
D6	1N4001	C1	2000 µF 25 V

Emergency lighting option

D1	1N5401	Transformer	12 V AC 4 A 50 VA
D2	1N5401	F1	1 A and fuseholder
Lamps and lampholders as required (up to 50 VA)		F2	1A

Switching mains powered equipment

D1	1N4001	12 V relay	16 A single pole
D2	1N4001	F1	rating as required and fuseholder

Chapter Four: full security house alarm

Digital control panel
12 V 2 ah battery
Fused spur, fused at 2 A (unswitched)

External alarm box (colour to suit)
118 dB electronic siren
12 V beacon strobe (colour to suit)
Self-activating bell (SAB) unit
Tamper circuit microswitch
Internal sounder

Seven passive IR sensors
Five vibration sensors
Three flush or surface door contacts
One garage door contact
Three personal attack buttons

200 m four-core alarm cable
100 m six-core alarm cable
Self-adhesive, 25 × 16 mm cable trunking as required
Mains cable type 6242Y for panel supply
3.5 mm cable clips, white, round (four-core)
4.0 mm cable clips, white, round (six-core)
7 mm cable clips flat, grey (mains cable)
One box 8 × 1.5 countersunk woodscrews
One box 8 × 1 countersunk woodscrews
One box of red plastic wall plugs
One strip of 2 A wire connectors

Chapter Five: testing and maintenance

DATE OF INSTALLATION:						
ZONE	AREA PROTECTED	PROTECTION DEVICES USED	TEST VALUE Ω	1st YEAR	2nd YEAR	3rd YEAR
1						
2						
3						
4						
5						
6						
7						
8						
PA						
TAMPER						

Figure 5.1   Installation and maintenance log.

Chapter Seven: external lighting systems

Figure 7.7. Security lighting circuits

500 W enclosed floodlight
Fused, switched spur unit
Mounting box for spur unit
Programmable timeswitch
Single pole timer bypass light switch
External photocell switch
Twin and earth cable 6242Y
Three core and earth cable 6243Y
Earth sleeving
External PIR sensor
Alarm interface

Figure 7.8. Low power lighting sounder circuits

12 V relay 16 A single pole
6 to 12 V buzzer
Four-way mains wiring junction box
On−off switch (may be a standard lighting switch)
9 V battery
18 SWG enamelled copper wire
Form A (normally open) reed switch (this may be the switch part of a
pair of flush door contacts)

Figure 7.9. Sensor-controlled floodlight and switched sounder

R1	1K
C1	1000 µF 25 V
LED1	red 5 mm
F1	10 mA
F2	1 A

12 V siren or sounder as required
Bridge rectifier 1 A 50 V
Transformer 9 V AC 1 A 20 VA
12 V relay 16 A single pole
Suitable enclosure
Switched, fused spur unit and mounting box
Four-way mains wiring junction box
Three gang two-way mains lighting switch
PVC mains installation cable 1 mm type 6242Y (two-core and earth)
    and 6243Y (three-core and earth)
Earth sleeving

Figure 7.10. A two-lamp, two-sensor system

Relay 1	240c AC two pole 5 A
Relay 2	240c AC two pole 5 A
switch 1 and 2	standard lighting switches both single pole two-way

Chapter Eight: Door entry systems

Figure 8.8. A simple intercom system

Telephone handset complete with carbon microphone and earpiece
Magnetic microphone
3 inch diameter 8 Ω loudspeaker
12 V DC buzzer
Plastic enclosure
Momentary pushbutton
1N4001 diode
100 m four-core alarm cable

Amplifier components

R1	1 m	C1	10 µF 16 V
R2	10K	C2	10 µF 16 V
R3	470R	C3	0.1 µF
R4	10R	C4	120 nF
PR1	10K linear preset	C5	10 µF 16 V
Q1	BC109	C6	100 µF 16 V
D1	1N4001	C7	220 µF 16 V
IC1	LM386	C8	56 nF
PCB to suit			

Chapter Nine. Stand-alone alarms

Figure 9.1. Door or window alarms

Circuit one: piezo transducer type

R1	1K
R2	680K
R3	680K
D1	1N4001
D2	1N4001
IC1	CD40106 hex Schmitt trigger

Suitable enclosure
9 V Battery
PCB as required
3 kHz transducer

C1	10 µF 16 V
C2	1 µF 63 V
C3	330 pF
SW1	Set of door contacts
SW2	Key switch

Circuit two: buzzer sounder

R1	100K
R2	680K
R3	33K
D1	1N4001
SW1	Door contacts
SW2	Key switch

9 V battery
PCB as required

C1	1 µF 63 V
C2	10 µF 16 V
IC1	CD4093
Q1	BC547
B1	6 V–12 V buzzer

Chapter Ten: Tools and equipment

Figure 10.2. Simple DC low voltage tester

R1	1K	D1	1N4001
R2	33K	ZD1	11 V Zener 400 mW
			dual red & green or
R3	1K	LED1	two single LEDs
Small plastic case			

Index

Printed in the United Kingdom
by Lightning Source UK Ltd.
131780UK00001B/62/A